陈恒 主编 / 楼偶俊 巩庆志 张立杰 副主编

Spring MVC
开发技术指南
微课版

清华大学出版社
北京

内 容 简 介

Spring MVC 是一款优秀的基于 MVC 思想的应用框架,是 Spring 的一个子框架。本书以大量实例介绍 Spring MVC + MyBatis 框架的基本思想、方法和技术,同时配备相应的实践环节,巩固 Spring MVC 应用开发的方法和技术,力图达到"做中学,学中做"。

全书共分为 14 章,内容包括 Spring 基础、Spring MVC 入门、Spring MVC 的 Controller、类型转换和格式化、数据绑定和表单标签库、拦截器、数据验证、国际化、统一异常处理、文件的上传和下载、EL 与 JSTL、MyBatis 入门、MyBatis 的映射器以及基于 Spring MVC + MyBatis 框架的名片管理系统的设计与实现等重要内容。书中实例以 Maven 管理项目依赖,侧重实用性和启发性,趣味性强、通俗易懂,使读者能够快速掌握 Spring MVC + MyBatis 框架的基础知识、编程技巧以及完整的开发体系,为适应实战应用打下坚实的基础。

本书既可作为大学计算机及相关专业的教材或教学参考书,也适合作为 Spring MVC 应用开发人员的参考用书。

本书封面贴有清华大学出版社防伪标签,无标签者不得销售。
版权所有,侵权必究。举报:010-62782989,beiqinquan@tup.tsinghua.edu.cn。

图书在版编目(CIP)数据

Spring MVC 开发技术指南:微课版/陈恒主编.—北京:清华大学出版社,2020.7(2024.8重印)
ISBN 978-7-302-55520-9

Ⅰ. ①S… Ⅱ. ①陈… Ⅲ. ①JAVA 语言-程序设计-指南 Ⅳ. ①TP312.8-62

中国版本图书馆 CIP 数据核字(2020)第 086394 号

责任编辑:张 玥
封面设计:常雪影
责任校对:白 蕾
责任印制:曹婉颖

出版发行:清华大学出版社
网　　址:https://www.tup.com.cn,https://www.wqxuetang.com
地　　址:北京清华大学学研大厦 A 座　　邮　编:100084
社 总 机:010-83470000　　邮　购:010-62786544
投稿与读者服务:010-62776969,c-service@tup.tsinghua.edu.cn
质量反馈:010-62772015,zhiliang@tup.tsinghua.edu.cn
课件下载:https://www.tup.com.cn,010-83470236

印 装 者:三河市铭诚印务有限公司
经　　销:全国新华书店
开　　本:185mm×260mm　　印 张:24　　字 数:555 千字
版　　次:2020 年 8 月第 1 版　　印 次:2024 年 8 月第 5 次印刷
定　　价:69.50 元

产品编号:087983-01

前 言
PREFACE

目前,尽管有许多与 Spring 框架有关的书籍,但国内单独介绍 Spring MVC + MyBatis 框架的书籍还寥寥无几。而且相关书籍非常注重知识的系统性,使得知识体系结构过于全面、庞大,这类书籍不太适合作为高校计算机相关专业的教材。同时,许多教师非常希望教材本身能引导学生尽可能地参与到教学活动中,因此本书的重点不是简单地介绍 Spring MVC + MyBatis 框架的基础知识,而是给出大量的实例与实践环节。通过学习本书,读者可以快速掌握 Spring MVC + MyBatis 框架,提高 Java Web 应用的开发能力。全书共 14 章,其各章的具体内容如下:

第 1 章 讲解 Spring 框架的基础知识,包括 Spring IoC、Spring AOP、Spring Bean 以及 Spring 的数据库编程及事务处理机制。

第 2 章 讲解 MVC 的设计思想以及 Spring MVC 开发环境的构建,同时还介绍了第一个 Spring MVC 应用的开发流程。

第 3 章 详细讲解基于注解的控制器、Controller 接收请求参数的方式以及编写请求处理方法,是本书的重点内容之一。

第 4 章 介绍类型转换器和格式化转换器,包括内置的类型转换器和格式化转换器以及自定义类型转换器和格式化转换器。

第 5 章 讲解数据绑定的基本原理、表单标签库的使用方法以及 JSON 数据交互,是本书的重点内容之一。

第 6 章 详细讲解 Spring MVC 框架中拦截器的概念、原理以及应用。

第 7 章 详细讲解 Spring MVC 框架的输入验证体系,包括 Spring 验证和 JSR 303 验证,是本书的重点内容之一。

第 8 章 介绍 Spring MVC 国际化的实现方法,包括 Java 国际化的思想、Spring MVC 的国际化以及用户自定义切换语言。

第 9 章 详细讲解使用 Spring MVC 框架进行异常统一处理的方法,是本书的重点内容之一。

第 10 章 讲解如何使用 Spring MVC 框架进行文件的上传与下载。

第 11 章 介绍 EL 与 JSTL 的基本用法。

第 12 章 讲解 MyBatis 的环境构建、工作原理以及与 Spring MVC 框架的整合开发,是本书的重点内容之一。

第 13 章 讲解 MyBatis 的 SQL 映射文件,包括 MyBatis 核心配置文件、SQL 映射文件、级联查询以及动态 SQL,是本书的重点内容之一。

第14章 通过名片管理系统的设计与实现,讲述使用Spring MVC ＋ MyBatis框架实现一个Web应用的方法,是本书的重点内容之一。

本书特别注重引导学生参与课堂教学活动,既适合作为大学计算机及相关专业的教材或教学参考书,也适合作为Spring MVC应用开发人员的参考用书。

为便于学习,本书提供教学课件、教学大纲、电子教案、程序源码以及实践环节与课后习题的参考答案,读者可登录清华大学出版社网站下载使用。另外录制了1000分钟微课视频,读书可在书中扫描二维码获取资源。

由于编者水平有限,书中难免会有不足之处,敬请广大读者批评指正。

编 者

2020年3月

目 录
CONTENTS

第1章 Spring 基础 ·· 1
- 1.1 Spring 概述 ·· 2
  - 1.1.1 Spring 的由来 ·· 2
  - 1.1.2 Spring 的体系结构 ·· 2
- 1.2 Spring 开发环境的构建 ·· 4
  - 1.2.1 使用 Eclipse 开发 Java Web 应用 ·· 4
  - 1.2.2 使用 STS（Spring Tool Suite）开发 Java Web 应用 ·· 8
  - 1.2.3 Spring 的下载及目录结构 ·· 9
  - 1.2.4 第一个 Spring 入门程序 ·· 10
  - 1.2.5 实践环节 ·· 12
- 1.3 Maven 管理 Spring 应用 ·· 12
  - 1.3.1 Maven 简介 ·· 12
  - 1.3.2 Maven 的 pom.xml ·· 13
  - 1.3.3 在 STS 中创建 Maven Web 项目 ·· 14
  - 1.3.4 使用 Maven 管理第一个 Spring 入门程序 ·· 19
  - 1.3.5 实践环节 ·· 21
- 1.4 Spring IoC ·· 21
  - 1.4.1 基本概念 ·· 21
  - 1.4.2 Spring 的常用注解 ·· 22
  - 1.4.3 基于注解的依赖注入 ·· 23
  - 1.4.4 Java 配置 ·· 26
  - 1.4.5 实践环节 ·· 29
- 1.5 Spring AOP ·· 29
  - 1.5.1 Spring AOP 的基本概念 ·· 29
  - 1.5.2 基于注解开发 AspectJ ·· 31
- 1.6 Spring Bean ·· 38
  - 1.6.1 Bean 的实例化 ·· 38

		1.6.2	Bean 的作用域 ………………………………………………	40

        1.6.2 Bean 的作用域 …………………………………………………… 40
        1.6.3 Bean 的初始化和销毁 …………………………………………… 43
  1.7 Spring 的数据库编程 ……………………………………………………… 45
        1.7.1 Spring JDBC 的 XML 配置 …………………………………… 45
        1.7.2 Spring JDBC 的 Java 配置 …………………………………… 46
        1.7.3 Spring JdbcTemplate 的常用方法 …………………………… 47
        1.7.4 基于@Transactional 注解的声明式事务管理 ……………… 54
        1.7.5 在事务处理中捕获异常 ………………………………………… 58
        1.7.6 实践环节 ………………………………………………………… 59
  1.8 本章小结 …………………………………………………………………… 59
  习题 1 …………………………………………………………………………… 59

第 2 章 Spring MVC 入门 …………………………………………………………… 60
  2.1 MVC 模式与 Spring MVC 工作原理 …………………………………… 61
        2.1.1 MVC 模式 ……………………………………………………… 61
        2.1.2 Spring MVC 工作原理 ………………………………………… 61
        2.1.3 Spring MVC 接口 ……………………………………………… 62
  2.2 第一个 Spring MVC 应用 ………………………………………………… 63
        2.2.1 创建 Maven 项目并添加依赖的 JAR 包 …………………… 63
        2.2.2 在 web.xml 文件中部署 DispatcherServlet ………………… 64
        2.2.3 创建 Web 应用首页 …………………………………………… 65
        2.2.4 创建 Controller 类 ……………………………………………… 65
        2.2.5 创建 Spring MVC 配置文件 ………………………………… 66
        2.2.6 应用的其他页面 ……………………………………………… 67
        2.2.7 发布并运行 Spring MVC 应用 ……………………………… 67
  2.3 基于 Java 配置的 Spring MVC 应用 …………………………………… 68
  2.4 实践环节 …………………………………………………………………… 70
  2.5 本章小结 …………………………………………………………………… 70
  习题 2 …………………………………………………………………………… 70

第 3 章 Spring MVC 的 Controller ………………………………………………… 71
  3.1 基于注解的控制器 ………………………………………………………… 72
        3.1.1 @Controller 注解类型 ………………………………………… 72
        3.1.2 @RequestMapping 注解类型 ………………………………… 72
        3.1.3 编写请求处理方法 …………………………………………… 73
  3.2 Controller 接收请求参数的常见方式 …………………………………… 75
        3.2.1 通过实体 Bean 接收请求参数 ……………………………… 75
        3.2.2 通过处理方法的形参接收请求参数 ………………………… 82

		3.2.3 通过 HttpServletRequest 接收请求参数 ………………………… 83
		3.2.4 通过@PathVariable 接收 URL 中的请求参数 …………………… 83
		3.2.5 通过@RequestParam 接收请求参数 …………………………… 84
		3.2.6 通过@ModelAttribute 接收请求参数 …………………………… 85

- 3.3 重定向与转发 ………………………………………………………………… 85
- 3.4 应用@Autowired 进行依赖注入 …………………………………………… 87
- 3.5 @ModelAttribute ……………………………………………………………… 89
- 3.6 实践环节 ……………………………………………………………………… 91
- 3.7 本章小结 ……………………………………………………………………… 91

习题 3 ………………………………………………………………………………… 91

第 4 章 类型转换和格式化 ……………………………………………………………… 92

- 4.1 类型转换的意义 ……………………………………………………………… 93
- 4.2 Converter ……………………………………………………………………… 94
  - 4.2.1 内置的类型转换器 ………………………………………………… 95
  - 4.2.2 自定义类型转换器 ………………………………………………… 96
  - 4.2.3 实践环节 …………………………………………………………… 102
- 4.3 Formatter ……………………………………………………………………… 102
  - 4.3.1 内置的格式化转换器 ……………………………………………… 102
  - 4.3.2 自定义格式化转换器 ……………………………………………… 102
  - 4.3.3 实践环节 …………………………………………………………… 108
- 4.4 本章小结 ……………………………………………………………………… 108

习题 4 ………………………………………………………………………………… 108

第 5 章 数据绑定和表单标签库 ………………………………………………………… 109

- 5.1 数据绑定 ……………………………………………………………………… 110
- 5.2 Spring 的表单标签库 ………………………………………………………… 110
  - 5.2.1 表单标签 …………………………………………………………… 110
  - 5.2.2 input 标签 ………………………………………………………… 111
  - 5.2.3 password 标签 …………………………………………………… 111
  - 5.2.4 hidden 标签 ……………………………………………………… 111
  - 5.2.5 textarea 标签 …………………………………………………… 112
  - 5.2.6 checkbox 标签 …………………………………………………… 112
  - 5.2.7 checkboxes 标签 ………………………………………………… 112
  - 5.2.8 radiobutton 标签 ………………………………………………… 113
  - 5.2.9 radiobuttons 标签 ………………………………………………… 113
  - 5.2.10 select 标签 ……………………………………………………… 113
  - 5.2.11 options 标签 …………………………………………………… 113

	5.2.12 errors 标签 ··· 113
5.3	数据绑定应用 ·· 114
	5.3.1 创建 Maven 项目并添加相关依赖 ·· 114
	5.3.2 Spring MVC 及 Web 相关配置 ·· 115
	5.3.3 领域模型 ··· 116
	5.3.4 Service 层 ·· 117
	5.3.5 Controller 层 ·· 117
	5.3.6 View 层 ·· 119
	5.3.7 测试应用 ··· 122
5.4	实践环节 ··· 123
5.5	JSON 数据交互 ··· 123
	5.5.1 JSON 概述 ·· 123
	5.5.2 JSON 数据转换 ·· 125
5.6	本章小结 ··· 132

习题 5 ·· 132

第 6 章 拦截器 ··· 133

6.1	拦截器概述 ·· 134
	6.1.1 拦截器的定义 ·· 134
	6.1.2 拦截器的配置 ·· 135
6.2	拦截器的执行流程 ·· 136
	6.2.1 单个拦截器的执行流程 ··· 136
	6.2.2 多个拦截器的执行流程 ··· 140
6.3	应用案例——用户登录权限验证 ·· 144
6.4	本章小结 ··· 151

习题 6 ·· 151

第 7 章 数据验证 ·· 152

7.1	数据验证概述 ·· 153
	7.1.1 客户端验证 ·· 153
	7.1.2 服务器端验证 ·· 153
7.2	Spring 验证器 ·· 153
	7.2.1 Validator 接口 ·· 153
	7.2.2 ValidationUtils 类 ··· 154
	7.2.3 验证示例 ··· 154
	7.2.4 实践环节 ··· 165
7.3	JSR 303 验证 ··· 166
	7.3.1 JSR 303 验证配置 ·· 166

7.3.2　标注类型 …………………………………………………………… 166
　　　7.3.3　验证示例 …………………………………………………………… 168
　　　7.3.4　实践环节 …………………………………………………………… 172
　7.4　本章小结 ………………………………………………………………………… 172
　习题 7 ………………………………………………………………………………… 172

第 8 章　国际化 …………………………………………………………………………… 173
　8.1　程序国际化概述 ………………………………………………………………… 174
　　　8.1.1　Java 国际化的思想 …………………………………………………… 174
　　　8.1.2　Java 支持的语言和国家 ……………………………………………… 174
　　　8.1.3　Java 程序国际化 ……………………………………………………… 175
　　　8.1.4　带占位符的国际化信息 ……………………………………………… 176
　　　8.1.5　实践环节 ……………………………………………………………… 177
　8.2　Spring MVC 的国际化 …………………………………………………………… 177
　　　8.2.1　Spring MVC 加载资源属性文件 ……………………………………… 178
　　　8.2.2　语言区域的选择 ……………………………………………………… 178
　　　8.2.3　使用 message 标签显示国际化信息 ………………………………… 179
　8.3　用户自定义切换语言示例 ……………………………………………………… 180
　8.4　本章小结 ………………………………………………………………………… 187
　习题 8 ………………………………………………………………………………… 187

第 9 章　统一异常处理 …………………………………………………………………… 188
　9.1　示例介绍 ………………………………………………………………………… 189
　9.2　SimpleMappingExceptionResolver 类 ………………………………………… 196
　9.3　HandlerExceptionResolver 接口 ……………………………………………… 197
　9.4　@ExceptionHandler 注解 ……………………………………………………… 199
　9.5　@ControllerAdvice 注解 ……………………………………………………… 200
　9.6　本章小结 ………………………………………………………………………… 201
　习题 9 ………………………………………………………………………………… 201

第 10 章　文件的上传和下载 …………………………………………………………… 202
　10.1　文件上传 ………………………………………………………………………… 203
　　　10.1.1　commons-fileupload 组件 ………………………………………… 203
　　　10.1.2　基于表单的文件上传 ……………………………………………… 203
　　　10.1.3　MultipartFile 接口 ………………………………………………… 204
　　　10.1.4　单文件上传 ………………………………………………………… 204
　　　10.1.5　多文件上传 ………………………………………………………… 211
　　　10.1.6　实践环节 …………………………………………………………… 216

10.2 文件下载 ·· 217
　　10.2.1 文件下载的实现方法 ·· 217
　　10.2.2 文件下载 ·· 217
10.3 本章小结 ·· 222
习题 10 ·· 222

第 11 章 EL 与 JSTL ·· 223

11.1 表达式语言 EL ·· 224
　　11.1.1 基本语法 ·· 224
　　11.1.2 EL 隐含对象 ·· 226
　　11.1.3 实践环节 ·· 231
11.2 JSP 标准标签库 JSTL ·· 232
　　11.2.1 配置 JSTL ·· 232
　　11.2.2 核心标签库之通用标签 ·· 233
　　11.2.3 核心标签库之流程控制标签 ·· 234
　　11.2.4 核心标签库之迭代标签 ·· 236
　　11.2.5 函数标签库 ·· 239
　　11.2.6 实践环节 ·· 242
11.3 本章小结 ·· 243
习题 11 ·· 243

第 12 章 MyBatis 入门 ·· 245

12.1 MyBatis 简介 ··· 246
12.2 MyBatis 的环境构建 ··· 246
　　12.2.1 非 Maven 构建 ·· 246
　　12.2.2 Maven 构建 ·· 246
12.3 MyBatis 的工作原理 ··· 247
12.4 使用 STS 开发 MyBatis 入门程序 ··· 248
　　12.4.1 创建 Maven 项目并添加相关依赖 ··· 248
　　12.4.2 创建 Log4j 的日志配置文件 ·· 249
　　12.4.3 创建持久化类 ·· 250
　　12.4.4 创建 SQL 映射文件 ·· 250
　　12.4.5 创建 MyBatis 的核心配置文件 ·· 251
　　12.4.6 创建测试类 ·· 252
12.5 MyBatis 与 Spring MVC 的整合开发 ·· 254
　　12.5.1 相关依赖 ·· 254
　　12.5.2 在 Sping MVC 的配置类中配置数据源及 MyBatis 工厂 ··················· 256
　　12.5.3 整合示例 ·· 258

	12.5.4 实践环节	263
12.6	使用 MyBatis Generator 插件自动生成映射文件	264
12.7	小结	266

习题 12 266

第 13 章 MyBatis 的映射器 267

13.1	MyBatis 的核心配置	268
13.2	映射器概述	268
13.3	<select>元素	269
	13.3.1 使用 Map 接口传递参数	270
	13.3.2 使用 Java Bean 传递参数	279
	13.3.3 使用@Param 注解传递参数	281
	13.3.4 <resultMap>元素	282
	13.3.5 使用 POJO 存储结果集	282
	13.3.6 使用 Map 存储结果集	285
	13.3.7 实践环节	287
13.4	<insert>元素	287
	13.4.1 主键（自动递增）回填	287
	13.4.2 自定义主键	290
13.5	<update>与<delete>元素	290
13.6	<sql>元素	290
13.7	级联查询	291
	13.7.1 一对一级联查询	291
	13.7.2 一对多级联查询	298
	13.7.3 多对多级联查询	304
13.8	动态 SQL	307
	13.8.1 <if>元素	307
	13.8.2 <choose><when><otherwise>元素	308
	13.8.3 <trim>元素	310
	13.8.4 <where>元素	311
	13.8.5 <set>元素	312
	13.8.6 <foreach>元素	314
	13.8.7 <bind>元素	315
13.9	本章小结	316

习题 13 316

第 14 章 名片管理系统的设计与实现 317

| 14.1 | 系统设计 | 318 |

14.1.1　系统功能需求 …………………………………… 318
　　　14.1.2　系统模块划分 …………………………………… 318
　14.2　数据库设计 …………………………………………… 318
　　　14.2.1　数据库概念结构设计 ……………………………… 318
　　　14.2.2　数据库逻辑结构设计 ……………………………… 319
　14.3　系统管理 ……………………………………………… 320
　　　14.3.1　Maven项目依赖管理 ……………………………… 320
　　　14.3.2　JSP页面管理 ……………………………………… 321
　　　14.3.3　包管理 …………………………………………… 323
　　　14.3.4　配置类管理 ……………………………………… 324
　　　14.3.5　配置文件管理 …………………………………… 329
　14.4　组件设计 ……………………………………………… 330
　　　14.4.1　工具类 …………………………………………… 330
　　　14.4.2　统一异常处理 …………………………………… 331
　　　14.4.3　验证码 …………………………………………… 332
　14.5　名片管理 ……………………………………………… 335
　　　14.5.1　领域模型与持久化类 ……………………………… 335
　　　14.5.2　Controller实现 …………………………………… 336
　　　14.5.3　Service实现 ……………………………………… 338
　　　14.5.4　Dao实现 ………………………………………… 342
　　　14.5.5　SQL映射文件 ……………………………………… 342
　　　14.5.6　添加名片 ………………………………………… 344
　　　14.5.7　查询名片 ………………………………………… 346
　　　14.5.8　修改名片 ………………………………………… 353
　　　14.5.9　删除名片 ………………………………………… 357
　14.6　用户相关 ……………………………………………… 357
　　　14.6.1　领域模型与持久化类 ……………………………… 357
　　　14.6.2　Controller实现 …………………………………… 358
　　　14.6.3　Service实现 ……………………………………… 359
　　　14.6.4　Dao实现 ………………………………………… 361
　　　14.6.5　SQL映射文件 ……………………………………… 361
　　　14.6.6　注册 ……………………………………………… 362
　　　14.6.7　登录 ……………………………………………… 364
　　　14.6.8　修改密码 ………………………………………… 367
　　　14.6.9　安全退出 ………………………………………… 368
　14.7　小结 …………………………………………………… 369
　习题14 ……………………………………………………… 369

参考文献 ………………………………………………………… 370

Spring 基础

学习目的与要求

本章重点讲解 Spring 的基础知识。通过本章的学习,了解 Spring 的体系结构,理解 Spring IoC 与 AOP 的基本原理,了解 Spring Bean 的生命周期、实例化以及作用域,掌握 Spring 的事务管理。

本章主要内容

- Spring 开发环境的构建。
- Spring IoC。
- Spring AOP。
- Spring Bean。
- Spring 的数据库编程。

　　Spring 是当前主流的 Java 开发框架,为企业级应用开发提供了丰富的功能。掌握 Spring 框架的使用,已是 Java 开发者必备的技能之一。本章将学习如何使用 Eclipse 或 STS(Spring Tool Suite)开发 Spring 程序,不过在此之前需要构建 Spring 的开发环境。

1.1 Spring 概述

1.1.1 Spring 的由来

Spring 是一个轻量级 Java 开发框架,最早由 Rod Johnson 创建,目的是为了解决企业级应用开发的业务逻辑层和其他各层的耦合问题。它是一个分层的 JavaSE/EEfull-stack(一站式)轻量级开源框架,为开发 Java 应用程序提供全面的基础架构支持。Spring 负责基础架构,因此 Java 开发者可以专注于应用程序的开发。

1.1.2 Spring 的体系结构

Spring 的功能模块被有组织地分散到约 20 个模块中,这些模块分布在核心容器 (Core Container)层、数据访问/集成(Data Access/Integration)层、Web 层、面向切面的编程(Aspect Oriented Programming,AOP)模块、植入(Instrumentation)模块、消息传输 (Messaging)和测试(Test)模块中,如图 1.1 所示。

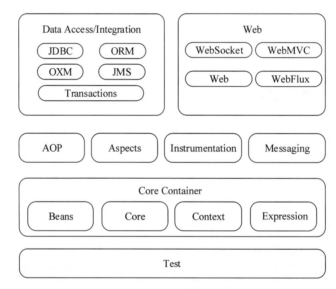

图 1.1 Spring 的体系结构

1. Core Container

Spring 的 Core Container 是其他模块建立的基础,由 Beans(spring-beans)、Core (spring-core)、Context(spring-context)和 Expression(spring-expression,Spring 表达式语言)等模块组成。

spring-beans 模块:提供了 BeanFactory,是工厂模式的一个经典实现,Spring 将管理对象称为 Bean。

spring-core 模块:提供了框架的基本组成部分,包括控制反转(Inversion of Control,

IoC)和依赖注入(Dependency Injection,DI)功能。

spring-context 模块：建立在 spring-beans 和 spring-core 模块基础上，提供一个框架式的对象访问方式，是访问定义和配置的任何对象媒介。

spring-expression 模块：提供了强大的表达式语言支持运行时查询和操作对象图。这是对 JSP 2.1 规范中规定的统一表达式语言(Unified EL)的扩展。该语言支持设置和获取属性值、属性分配、方法调用、访问数组、集合和索引器的内容、逻辑和算术运算、变量命名以及从 Spring 的 IoC 容器中以名称检索对象。它还支持列表投影、选择以及常见的列表聚合。

2. AOP 和 Instrumentation

Spring 框架中与 AOP 和 Instrumentation 相关的模块有 AOP(spring-aop)模块、Aspects(spring-aspects)模块以及 Instrumentation(spring-instrument)模块。

spring-aop 模块：提供了一个符合 AOP 要求的面向切面的编程实现，允许定义方法拦截器和切入点，将代码按照功能分离，以便干净地解耦。

spring-aspects 模块：提供了与 AspectJ 的集成功能，AspectJ 是一个功能强大且成熟的 AOP 框架。

spring-instrument 模块：提供了类植入(Instrumentation)支持和类加载器的实现，可以在特定的应用服务器中使用。Instrumentation 提供了一种虚拟机级别支持的 AOP 实现方式，使得开发者无须对 JDK 做任何升级和改动，就可以实现某些 AOP 的功能。

3. Messaging

Spring 4.0 以后新增了 Messaging(spring-messaging)模块，该模块提供了对消息传递体系结构和协议的支持。

4. Data Access/Integration

数据访问/集成层由 JDBC(spring-jdbc)、ORM(spring-orm)、OXM(spring-oxm)、JMS(spring-jms)和 Transactions(spring-tx)模块组成。

spring-jdbc 模块：提供了一个 JDBC 的抽象层，消除了烦琐的 JDBC 编码和数据库厂商特有的错误代码解析。

spring-orm 模块：为流行的对象关系映射(Object-Relational Mapping)API 提供集成层，包括 JPA 和 Hibernate。使用 spring-orm 模块，可以将这些 O/R 映射框架与 Spring 提供的所有其他功能结合使用，例如声明式事务管理功能。

spring-oxm 模块：提供了一个支持对象/XML 映射的抽象层实现，如 JAXB、Castor、JiBX 和 XStream。

spring-jms 模块(Java Messaging Service)：指 Java 消息传递服务，包含用于生产和使用消息的功能。自 Spring 4.1 后，提供了与 spring-messaging 模块的集成。

spring-tx 模块(事务模块)：支持用于实现特殊接口和所有 POJO(普通 Java 对象)类的编程和声明式事务管理。

5. Web

Web 层由 Web(spring-web)、WebMVC(spring-webmvc)、WebSocker(spring-websocket)和 WebFlux(spring-webflux)模块组成。

spring-web 模块：提供了基本的 Web 开发集成功能。例如：多文件上传功能、使用 Servlet 监听器初始化一个 IoC 容器以及 Web 应用上下文。

spring-webmvc 模块：也称为 Web-Servlet 模块，包含用于 Web 应用程序的 Spring MVC 和 REST Web Services 实现。Spring MVC 框架提供了领域模型代码和 Web 表单之间的清晰分离，并与 Spring Framework 的所有其他功能集成，本书后续章节将会详细讲解 Spring MVC 框架。

spring-websocket 模块：Spring 4.0 后新增的模块，它提供了 WebSocket 和 SockJS 的实现，主要是与 Web 前端的全双工通信的协议。

spring-webflux 是一个新的非堵塞函数式 Reactive Web 框架，可以用来建立异步的、非阻塞、事件驱动的服务，并且扩展性非常好(该模块是 Spring 5 的新增模块)。

6. Test

Test(spring-test)模块：支持使用 JUnit 或 TestNG 对 Spring 组件进行单元测试和集成测试。

1.2 Spring 开发环境的构建

使用 Spring 框架开发应用前，应先搭建开发环境。本书的开发环境都是基于 Eclipse 或 STS 平台的 Java Web 应用的开发环境。

1.2.1 使用 Eclipse 开发 Java Web 应用

为了提高开发效率，通常需要安装 IDE(集成开发环境)工具。Eclipse 是一个可用于开发 Web 应用的 IDE 工具。

登录 https://www.eclipse.org/，打开图 1.2 所示的 Eclipse 官方主页。

单击图 1.2 中的 Download 按钮，打开图 1.3 所示的 Eclipse IDE 2019-12 下载页面。

单击图 1.3 中的 Download Packages 超链接，打开图 1.4 所示的 Eclipse IDE 2019-12 版本选择页面。

根据操作系统的位数下载相应的 Eclipse。本书采用的 Eclipse 是 Windows 64-bit (文件名为 eclipse-jee-2019-12-R-win32-x86_64.zip)。

使用 Eclipse 之前，需要对 JDK、Web 服务器和 Eclipse 进行一些必要的配置。因此，安装 Eclipse 之前，应事先安装 JDK 和 Web 服务器。

1. 安装 JDK

登录 http://www.oracle.com/technetwork/java，选择 Java SE 提供的 JDK。根据操

图 1.2　Eclipse 官方主页

图 1.3　Eclipse IDE 2019-12 下载页面

图 1.4　Eclipse IDE 2019-12 版本选择页面

作系统的位数下载相应的 JDK。本书采用的 JDK 是 jdk-11.0.6_windows-x64_bin.exe。

按照提示安装完成 JDK 后,需要配置"环境变量"的"系统变量"Java_Home 和 Path。在 Windows 10 系统下,系统变量示例如图 1.5 和图 1.6 所示。

2. Web 服务器

目前,比较常用的 Web 服务器包括 Tomcat、JRun、Resin、WebSphere、WebLogic 等,本书采用的是 Tomcat 9.0。

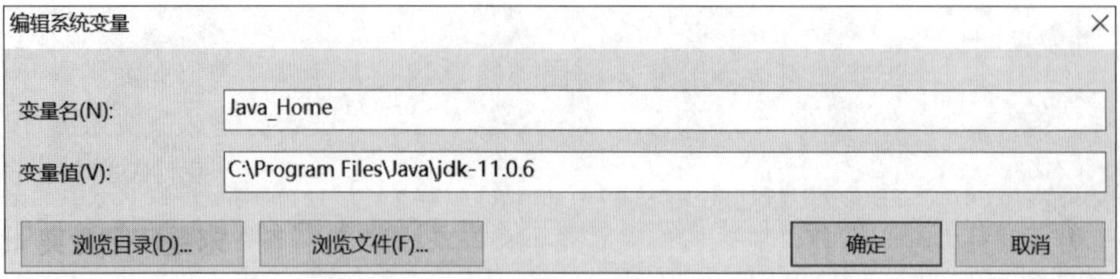

图 1.5 新建系统变量 Java_Home

图 1.6 编辑环境变量 Path

登录 Apache 软件基金会的官方网站 http://jakarta.Apache.org/tomcat，下载 Tomcat 9.0 的免安装版(本书采用的 Tomcat 是 apache-tomcat-9.0.30-windows-x64.zip)。登录网站后，首先在 Download 里选择 Tomcat 9，然后在 Binary Distributions 的 Core 中选择相应版本即可。

安装 Tomcat 之前需要事先安装 JDK 并配置系统环境变量 Java_Home。将下载的 apache-tomcat-9.0.30-windows-x64.zip 解压到某个目录下，比如解压到 E:\Java soft，解压缩后将出现图 1.7 所示的目录结构。

执行 Tomcat 根目录下 bin 文件夹中的 startup.bat 来启动 Tomcat 服务器。执行 startup.bat 启动 Tomcat 服务器会占用一个 MS-DOS 窗口，如果关闭当前 MS-DOS 窗口，将关闭 Tomcat 服务器。启动 Tomcat 服务器后，在浏览器的地址栏中输入 http://localhost:8080，将出现图 1.8 所示的 Tomcat 测试页面。

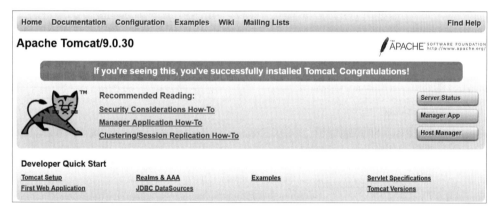

图 1.7　Tomcat 目录结构

图 1.8　Tomcat 测试页面

3. 安装 Eclipse

Eclipse 下载完成后，解压到自己设置的路径下，即可完成安装。安装 Eclipse 后，双击 Eclipse 安装目录下的 eclipse.exe 文件，启动 Eclipse。

4. 集成 Tomcat

启动 Eclipse，选择 Window/Preferences 菜单项，在弹出的对话框中选择 Server/Runtime Environments 命令。在弹出的窗口中单击 Add 按钮，弹出图 1.9 所示的 New Server Runtime Environment 界面，在此可以配置各种版本的 Web 服务器。

在图 1.9 中选择 Apache Tomcat v9.0 服务器版本，单击 Next 按钮，进入图 1.10 所示界面。

在图 1.10 中单击 Browse... 按钮，选择 Tomcat 的安装目录，单击 Finish 按钮即可完成 Tomcat 配置。

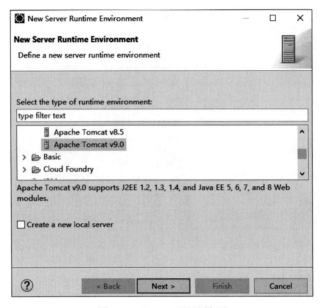

图 1.9　Tomcat 配置界面

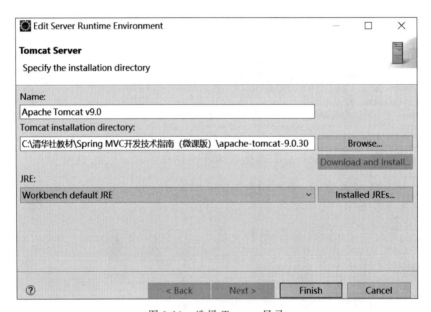

图 1.10　选择 Tomcat 目录

至此，可以使用 Eclipse 创建 Dynamic Web Project(Java Web 应用)，并在 Tomcat 下运行。

1.2.2　使用 STS(Spring Tool Suite)开发 Java Web 应用

STS 是一个专为 Spring 开发定制的 Eclipse，方便创建、调试、运行及维护 Spring 应用。使用该工具，可以很方便地生成一个 Spring 工程，比如 Web 工程。

可通过官方网站 https://spring.io/tools 下载 STS，本书采用的版本是 spring-tool-suite-4-4.5.1.RELEASE-e4.14.0-win32.win32.x86_64.self-extracting.jar。该版本与 Eclipse 一样，双击解压即完成安装。

与 Eclipse 一样，使用 STS 之前，需要对 JDK、Web 服务器和 STS 进行一些必要的配置。这些配置与 1.2.1 节完全一样，这里不再赘述。配置完成后，即可使用 STS 创建 Dynamic Web Project，并在 Tomcat 下运行。

读者既可以使用 Eclipse 平台测试本书代码，也可以使用 STS 测试本书代码。

1.2.3 Spring 的下载及目录结构

使用 Spring 框架开发应用程序时，除了引用 Spring 自身的 JAR 包外，还需要引用 commons.logging 的 JAR 包。

1. Spring 的 JAR 包

Spring 官方网站升级后，建议通过 Maven 和 Gradle 下载。对于不使用 Maven 和 Gradle 下载的开发者，本书给出一个 Spring Framework JAR 官方直接下载路径：http://repo.springsource.org/libs-release-local/org/springframework/spring/。本书采用的是 spring-5.2.3.RELEASE-dist.zip。将下载到的 ZIP 文件解压缩，解压缩后的目录结构如图 1.11 所示。

名称	修改日期	类型	大小
docs	2020/1/14 7:59	文件夹	
libs	2020/1/14 7:59	文件夹	
schema	2020/1/14 7:59	文件夹	
license.txt	2020/1/14 7:50	文本文档	15 KB
notice.txt	2020/1/14 7:50	文本文档	1 KB
readme.txt	2020/1/14 7:50	文本文档	1 KB

图 1.11 springframework-5.2.3 的目录结构

在图 1.11 中，docs 目录包含 Spring 的 API 文档和开发规范。

在图 1.11 中，libs 目录包含开发 Spring 应用需要的 JAR 包和源代码。该目录下有三类 JAR 文件，其中，以 RELEASE.jar 结尾的文件是 Spring 框架 class 的 JAR 包，即开发 Spring 应用所需要的 JAR 包；以 RELEASE-javadoc.jar 结尾的文件是 Spring 框架 API 文档的压缩包；以 RELEASE-sources.jar 结尾的文件是 Spring 框架源文件的压缩包。libs 目录中有四个基础包：spring-core-5.2.3.RELEASE.jar、spring-beans-5.2.3.RELEASE.jar、spring-context-5.2.3.RELEASE.jar 和 spring-expression-5.2.3.RELEASE.jar，分别对应 Spring 核心容器的四个模块：Spring-core 模块、Spring-beans 模块、Spring-context 模块和 Spring-expression 模块。

在图 1.11 中，schema 目录包含开发 Spring 应用需要的 schema 文件，这些 schema 文件定义了 Spring 相关配置文件的约束。

2. commons.logging 的 JAR 包

Spring 框架依赖于 Apache Commons Logging 组件，该组件的 JAR 包可以通过官方网站 http://commons.apache.org/proper/commons-logging/download_logging.cgi 下载，本书下载的是 commons-logging-1.2-bin.zip，解压缩后，即可找到 commons-logging-1.2.jar。

对于 Spring 框架的初学者，开发 Spring 应用时，只需要将 Spring 的四个基础包和 commons-logging-1.2.jar 复制到 Web 应用的 WEB-INF/lib 目录下。如果不明白需要哪些 JAR 包，可以将 Spring 的 libs 目录中 spring-XXX-5.2.3.RELEASE.jar 全部复制到 WEB-INF/lib 目录下。

1.2.4 第一个 Spring 入门程序

本节通过一个简单的入门程序演示 Spring 框架的使用过程。

【例 1-1】 Spring 框架的使用过程。

具体实现步骤如下。

1. 使用 STS 创建 Web 应用，并导入 JAR 包

使用 STS 创建一个名为 ch1_1 的 Dynamic Web Project，并将 Spring 的四个基础包和第三方依赖包 commons-logging-1.2.jar 复制到 ch1_1 的 WEB-INF/lib 目录中，如图 1.12 所示。

图 1.12 Web 应用 ch1_1 导入的 JAR 包

注意：讲解 Spring MVC 框架前，本书的实例并没有真正运行 Web 应用。创建 Web 应用的目的是方便添加相关 JAR 包。

2. 创建接口 TestDao

Spring 解决的是业务逻辑层和其他各层的耦合问题,因此它将面向接口的编程思想贯穿整个系统应用。

在 src 目录下创建一个 dao 包,并在 dao 包中创建接口 TestDao,接口中定义一个 sayHello()方法,代码如下:

```
package dao;
public interface TestDao {
    public void sayHello();
}
```

3. 创建接口 TestDao 的实现类 TestDaoImpl

在 dao 包下创建接口 TestDao 的实现类 TestDaoImpl,代码如下:

```
package dao;
public class TestDaoImpl implements TestDao{
    @Override
    public void sayHello() {
        System.out.println("Hello, Study hard!");
    }
}
```

4. 创建配置文件 applicationContext.xml

在 src 目录下创建 Spring 的配置文件 applicationContext.xml,并在该文件中使用实现类 TestDaoImpl 创建一个 id 为 test 的 Bean,代码如下:

```xml
<?xml version="1.0" encoding="UTF-8"?>
<beans xmlns="http://www.springframework.org/schema/beans"
    xmlns:xsi="http://www.w3.org/2001/XMLSchema-instance"
    xsi:schemaLocation="http://www.springframework.org/schema/beans
        http://www.springframework.org/schema/beans/spring-beans.xsd">
    <!--将指定类 TestDaoImpl 配置给 Spring,让 Spring 创建其实例 -->
    <bean id="test" class="dao.TestDaoImpl" />
</beans>
```

注:配置文件的名称可以自定义,但习惯上命名为 applicationContext.xml,有时也命名为 beans.xml。配置文件信息不需要读者手写,可以从 Spring 的帮助文档中复制(首先使用浏览器打开\spring-framework-5.2.3.RELEASE\docs\spring-framework-reference\index.html,在页面中单击超链接 Core,在 1.2.1Configuration Metadata 小节下即可找到配置文件的约束信息)。

5. 创建测试类

在 src 目录下创建一个 test 包，并在 test 包中创建 Test 类，代码如下：

```
package test;
import org.springframework.context.ApplicationContext;
import org.springframework.context.support.ClassPathXmlApplicationContext;
import dao.TestDao;
public class Test {
    public static void main(String[] args) {
        //初始化 Spring 容器 ApplicationContext,加载配置文件
        //@SuppressWarnings 抑制警告的关键字,有泛型未指定类型
       @SuppressWarnings("resource")
        ApplicationContext appCon =new ClassPathXmlApplicationContext
("applicationContext.xml");
        //通过容器获取 test 实例
        TestDao tt =(TestDao)appCon.getBean("test");//test 为配置文件中的 id
        tt.sayHello();
    }
}
```

执行上述 main()方法后，将在控制台输出 Hello, Study hard!。上述 main()方法中并没有使用 new 运算符创建 TestDaoImpl 类的对象，而是通过 Spring 容器来获取实现类 TestDaoImpl 的对象，这就是 Spring IoC 的工作机制。

1.2.5 实践环节

(1) 下载最新版本的 Spring JAR 包。
(2) 在 STS 中搭建 Spring 的开发环境。

1.3　Maven 管理 Spring 应用

1.3.1 Maven 简介

　　Apache Maven 是一个软件项目管理工具。基于项目对象模型 (Project Object Model,POM)的理念,通过一段核心描述信息来管理项目构建、报告和文档信息。在 Java 项目中,Maven 主要完成两件工作：①统一开发规范与工具;②统一管理 JAR 包。

　　Maven 统一管理项目开发需要的 JAR 包,但这些 JAR 包将不再包含在项目内(即不在 lib 目录下),而是存放于仓库中。仓库主要包括如下内容：

1. 中央仓库

存放开发过程中的所有 JAR 包,例如 JUnit,可以通过互联网从中央仓库下载,下载

地址为 http://mvnrepository.com。

2. 本地仓库

本地计算机中的仓库。官方下载 Maven 的本地仓库,配置在%MAVEN_HOME%\conf\settings.xml 文件中,找到 localRepository 即可;Eclipse 或 STS 中自带 Maven 的默认本地仓库地址在{user.home}/.m2/repository/settings.xml 文件中,同样找到 localRepository 即可。

Maven 项目首先从本地仓库中获取需要的 JAR 包,当无法获取指定的 JAR 包时,本地仓库将从远程仓库(中央仓库)中下载 JAR 包,并放入本地仓库,以备将来使用。

1.3.2 Maven 的 pom.xml

Maven 是基于项目对象模型的理念管理项目的,所以 Maven 的项目都有一个 pom.xml 配置文件来管理项目的依赖以及项目的编译等功能。

在 Maven Web 项目中,重点关注以下元素。

1. properties 元素

在<properties></properties>之间可以定义变量,以便在<dependency></dependency>中引用,示例代码如下:

```xml
<properties>
    <!--spring 版本号 -->
    <spring.version>5.2.3.RELEASE</spring.version>
</properties>
<dependencies>
    <dependency>
        <groupId>org.springframework</groupId>
        <artifactId>spring-core</artifactId>
        <version>${spring.version}</version>
    </dependency>
</dependencies>
```

2. dependencies 元素

<dependencies></dependencies>,此元素包含多个项目依赖需要使用的<dependency></dependency>元素。

3. dependency 元素

<dependency></dependency>元素内部通过<groupId></groupId>、<artifactId></artifactId>、<version></version>三个子元素确定唯一的依赖,也可以称为三个坐标。示例代码如下:

```xml
<dependency>
    <!--groupId 组织的唯一标识   -->
    <groupId>org.springframework</groupId>
    <!--artifactId 项目的唯一标识   -->
    <artifactId>spring-core</artifactId>
    <!--version 项目的版本号 -->
    <version>${spring.version}</version>
</dependency>
```

4. scope 子元素

在<dependency></dependency>元素中,有时使用<scope></scope>子元素管理依赖的部署。<scope></scope>子元素可以使用以下 5 个值:

(1) compile(编译范围)。

compile 是缺省值,即默认范围。依赖如果没有提供范围,那么该依赖的范围就是编译范围。编译范围的依赖在所有的 classpath 中可用,同时也会被打包发布。

(2) provided(已提供范围)。

provided 表示已提供范围,只有当 JDK 或者容器已提供该依赖才可以使用。已提供范围的依赖不是传递性的,也不会被打包发布。

(3) runtime(运行时范围)。

runtime 范围依赖在运行和测试系统时需要,但在编译时不需要。

(4) test(测试范围)。

test 范围依赖在一般的编译和运行时都不需要,它们只有在测试编译和测试运行阶段可用。不会随项目发布。

(5) system(系统范围)。

system 范围与 provided 范围类似,但需要显式地提供包含依赖的 JAR 包,Maven 不会在 Repository 中查找它。

1.3.3 在 STS 中创建 Maven Web 项目

本节在基于 STS 平台的 Java Web 应用开发环境(1.2.2 节)的基础上创建 Maven 项目,具体步骤如下。

1. 新建 Maven 项目

在 STS 中新建 Maven 项目,具体步骤如下。

(1) 通过选择菜单中的 File→New→Project→Maven→Maven Project,打开图 1.13 所示的 Select project name and location 对话框。

(2) 在图 1.13 中勾选 Create a simple project 复选框,单击 Next 按钮,打开图 1.14 所示的 Maven 项目信息输入界面。

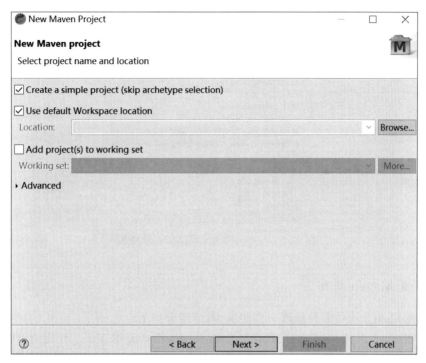

图 1.13　Select project name and location 对话框

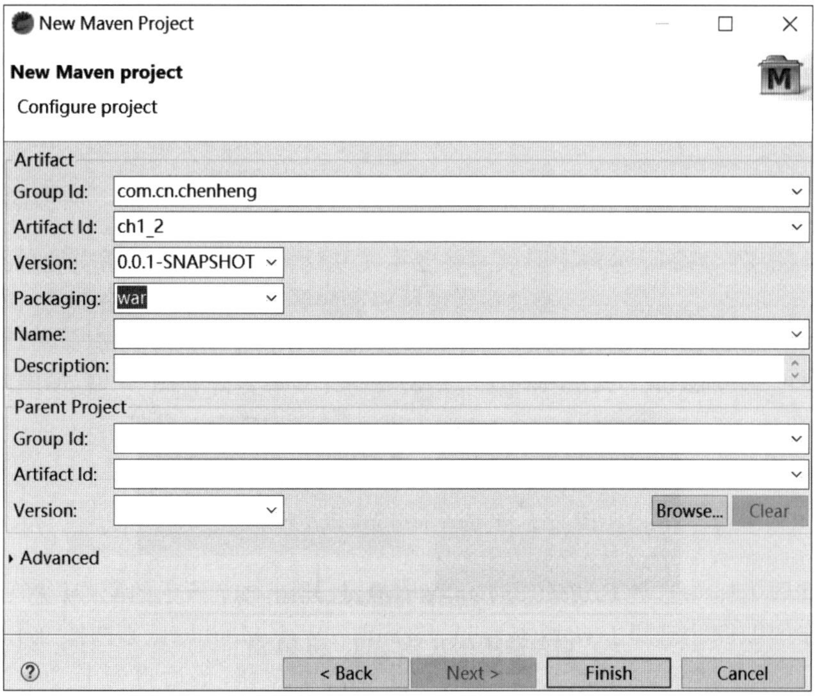

图 1.14　Maven 项目信息输入界面

（3）在图1.14中单击Finish按钮，完成Maven项目ch1_2的创建。ch1_2的目录结构如图1.15所示。

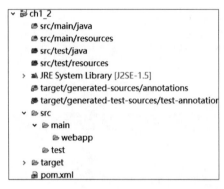

图1.15 Maven项目ch1_2的目录结构

2．配置Maven项目

新建的Maven项目需要修改一些配置，具体步骤如下。

（1）在ch1_2的项目名上右击，选择Properties菜单项，打开图1.16所示的Properties for ch1_2对话框。

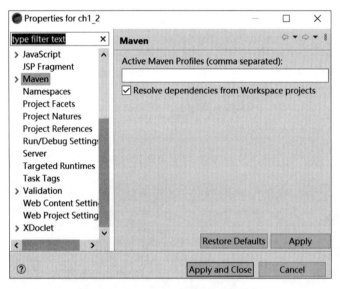

图1.16 Properties for ch1_2对话框

（2）在图1.16中选择Project Facets菜单项，打开图1.17所示的配置选择项界面。

（3）在图1.17中选择Java版本为11，Dynamic Web Module版本选择为4.0；然后勾掉Dynamic Web Module前的复选框，并单击Apply按钮；最后再勾上Dynamic Web Module前的复选框，出现Further configuration available…超链接，如图1.18所示。

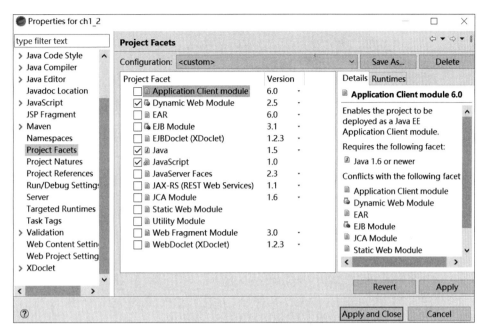

图 1.17 Project Facets 配置选择项界面

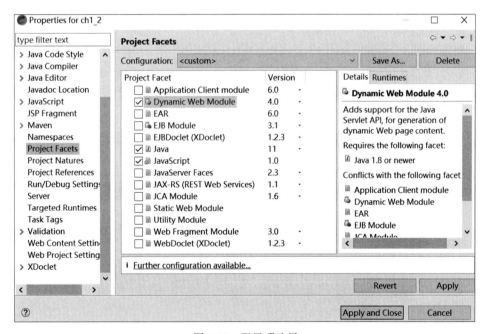

图 1.18 配置项选择

（4）在图 1.18 中单击 Further configuration available...超链接，打开图 1.19 所示的 Configure web module settings 界面。在该界面中，修改 Content directory 的内容为 src/main/webapp，并勾选图 1.19 中的复选框，然后单击 OK 按钮，返回图 1.18。然后单击

图 1.18 中的 Apply 按钮。

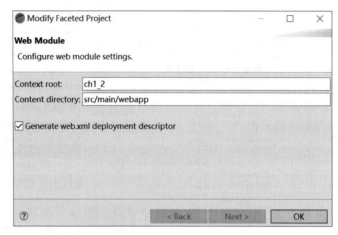

图 1.19　Configure web module settings 界面

（5）在图 1.18 中单击 Runtimes 选择项，并勾选 Dynamic Web Module 和 Apache Tomcat v 9.0 前的复选框，如图 1.20 所示。最后，单击图 1.20 中的 Apply and Close 按钮。至此，Maven 项目配置完毕。配置完毕的 Maven 项目 ch1_2 的目录结构如图 1.21 所示。

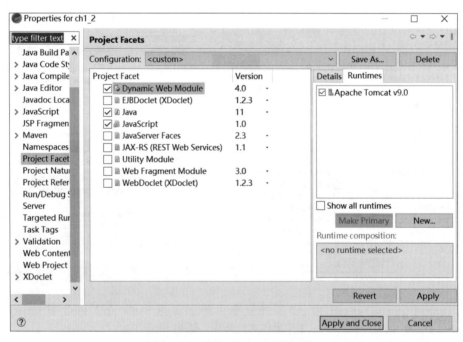

图 1.20　选择 Runtimes 选择项

图 1.21 中的 src/main/java 目录包含项目的 Java 源代码；src/main/resources 目录包含项目所需的资源（如配置文件）；src/test/java 目录包含用于测试的 Java 代码；src/

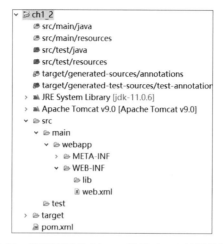

图 1.21　配置完毕的 Maven 项目 ch1_2 的目录结构

main/webapp 目录包含 Java Web 应用程序。

3. 测试 Maven 项目

首先,在项目 ch1_2 的 src/main/webapp 目录下创建名为 index.jsp 的页面。然后选中 index.jsp 的文件名,右击选择 Run As→Run on Server,测试运行 index.jsp 页面。运行结果如图 1.22 所示。

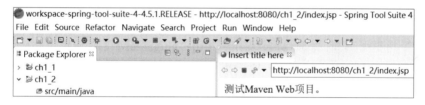

图 1.22　Maven 项目测试结果

1.3.4　使用 Maven 管理第一个 Spring 入门程序

本节在 1.3.3 节 Maven Web 项目 ch1_2 的基础上构建 1.2.4 节的第一个 Spring 入门程序。

【例 1-2】　使用 Maven 构建 Spring 入门程序。

具体步骤如下。

1. Maven 管理项目依赖的 JAR 包

这里,只需将【例 1-1】的第一步导入 JAR 包的方式修改为 Maven 管理的方式。即在项目 ch1_2 的 pom.xml 文件中,添加项目的相关依赖,pom.xml 文件的代码如下:

```
<project xmlns="http://maven.apache.org/POM/4.0.0"
    xmlns:xsi="http://www.w3.org/2001/XMLSchema-instance"
```

```xml
    xsi:schemaLocation="http://maven.apache.org/POM/4.0.0
https://maven.apache.org/xsd/maven-4.0.0.xsd">
    <modelVersion>4.0.0</modelVersion>
    <groupId>com.cn.chenheng</groupId>
    <artifactId>ch1_2</artifactId>
    <version>0.0.1-SNAPSHOT</version>
    <packaging>war</packaging>
    <properties>
        <!--spring 版本号 -->
        <spring.version>5.2.3.RELEASE</spring.version>
    </properties>
    <dependencies>
        <dependency>
            <groupId>org.springframework</groupId>
            <artifactId>spring-context</artifactId>
            <version>${spring.version}</version>
        </dependency>
    </dependencies>
</project>
```

从上述 pom.xml 文件内容可以看出只添加了 spring-context 依赖。这是因为 Maven 根据 spring-context 的依赖关系自动添加了相关依赖。

spring-context 定义了 Spring 容器,并依赖了 spring-core、spring-expression、spring-aop 和 spring-beans。依赖关系如图 1.23 所示。

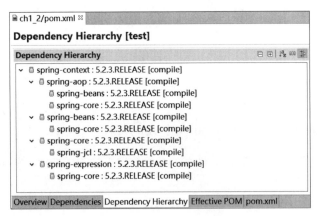

图 1.23 依赖关系图

从图 1.23 中可以看出 spring-core 依赖 spring-jcl(commons-logging)。所以,在 pom.xml 文件中也没有添加 commons-logging 依赖。

请记住 spring-context 的依赖关系,对 Spring 框架后续学习很重要。

2. 复制相关代码文件

将 ch1_1 src 目录下的 dao 包、test 包和 applicationContext.xml 配置文件复制到 ch1_2 的 src/main/java 目录下，即可进行代码测试。

【例 1-2】中使用 Maven 构建 Spring 入门程序。与【例 1-1】不同的是管理项目所依赖的 JAR 包的方式。【例 1-1】是将 Spring 的四个基础包和第三方依赖包 commons-logging-1.2.jar 复制到项目的 WEB-INF/lib 目录中。而在【例 1-2】中，是使用 pom.xml 文件管理项目所依赖的 JAR 包。

读者可根据自己的习惯来管理项目依赖的 JAR 包，但本书后续的程序都是使用 Maven 管理项目依赖的 JAR 包。

1.3.5　实践环节

参考 1.3.3 节，使用 STS 创建 Maven 项目 practice135，并配置与测试该 Maven 项目。

1.4　Spring IoC

1.4.1　基本概念

控制反转（IoC）是一个比较抽象的概念，是 Spring 框架的核心，用来消减计算机程序的耦合问题。依赖注入（DI）是 IoC 的另外一种说法，只是从不同的角度描述相同的概念。下面通过实际生活中的一个例子解释 IoC 和 DI。

人们需要一件东西时，第一反应就是找东西，例如想吃面包。没有面包店和有面包店两种情况下，您会怎么做？没有面包店时，最直观的做法可能是按照自己的口味制作面包，也就是需要主动制作一个面包。然而，时至今日，各种面包店很盛行，如果不想制作面包，可以把自己的口味告诉店家，一会儿就可以吃到可口的面包了。注意，这里并没有制作面包，而是由店家制作，但是完全符合自己的口味。

上面只是列举了一个非常简单的例子，但包含了控制反转的思想，即把制作面包的主动权交给店家。下面通过面向对象编程思想继续探讨这两个概念。

当某个 Java 对象（调用者，比如您）需要调用另一个 Java 对象（被调用者，即被依赖对象，比如面包）时，在传统编程模式下，调用者通常会采用"new 被调用者"的代码方式创建对象（比如您自己制作面包）。这种方式会增加调用者与被调用者之间的耦合性，不利于后期代码的升级与维护。

Spring 框架出现后，对象的实例不再由调用者创建，而是由 Spring 容器（比如面包店）创建。Spring 容器会负责控制程序之间的关系（比如面包店负责控制您与面包的关系），而不是由调用者的程序代码直接控制。这样，控制权由调用者转移到 Spring 容器，控制权发生了反转，这就是 Spring 的控制反转。

从 Spring 容器角度来看，Spring 容器负责将被依赖对象赋值给调用者的成员变量，相当于为调用者注入它所依赖的实例,这就是 Spring 的依赖注入,主要目的是为了解耦,体现一种"组合"的理念。

综上所述,控制反转是一种通过描述(在 Spring 中可以是 XML 或注解)并通过第三方去产生或获取特定对象的方式。在 Spring 中实现控制反转的是 IoC 容器,其实现方法是依赖注入。

1.4.2 Spring 的常用注解

在 Spring 框架中,尽管使用 XML 配置文件可以很简单地装配 Bean(如【例 1-1】),但如果应用中有大量的 Bean 需要装配时,会导致 XML 配置文件过于庞大,不方便以后的升级维护。因此,更多时候推荐开发者使用注解(annotation)的方式装配 Bean。

Spring 框架中定义了一系列的注解,常用注解如下。

1. 声明 Bean 的注解

(1) @Component。

该注解是一个泛化的概念,仅仅表示一个组件对象(Bean),可以作用在任何层次上,没有明确的角色。

(2) @Repository。

该注解用于将数据访问层(DAO)的类标识为 Bean,即注解数据访问层 Bean,功能与@Component()相同。

(3) @Service。

该注解用于标注一个业务逻辑组件类(Service 层),功能与@Component()相同。

(4) @Controller。

该注解用于标注一个控制器组件类(Spring MVC 的 Controller),功能与@Component()相同。

2. 注入 Bean 的注解

(1) @Autowired。

该注解可以对类成员变量、方法及构造方法进行标注,完成自动装配工作。通过@Autowired 的使用来消除 setter 和 getter 方法。默认按照 Bean 的类型进行装配。

(2) @Resource。

该注解与@Autowired 功能一样。区别在于,该注解默认是按照名称来装配注入的,只有找不到与名称匹配的 Bean 才会按照类型来装配注入;而@Autowired 默认按照 Bean 的类型进行装配,如果想按照名称来装配注入,则需要结合@Qualifier 注解一起使用。

@Resource 注解有两个属性:name 和 type。name 属性指定 Bean 的实例名称,即按照名称来装配注入;type 属性指定 Bean 类型,即按照 Bean 的类型进行装配。

(3) @Qualifier。

该注解与@Autowired 注解配合使用。当@Autowired 注解需要按照名称来装配注入，则需要结合该注解一起使用，Bean 的实例名称由@Qualifier 注解的参数指定。

1.4.3 基于注解的依赖注入

Spring IoC 容器负责创建和注入 Bean。Spring 提供使用 XML 配置、注解、Java 配置以及 groovy 配置实现 Bean 的创建和注入。本书尽量使用注解（@Component、@Repository、@Service 以及@Controller 等业务 Bean 的配置）和 Java 配置（全局配置如数据库、MVC 等相关配置）完全代替 XML 配置，这也是 Spring Boot 推荐的配置方式。

下面通过一个简单实例演示基于注解的依赖注入的使用过程。

【例 1-3】 基于注解的依赖注入的使用过程。该实例的具体要求是：在 Controller 层中依赖注入 Service 层；在 Service 层中依赖注入 DAO 层。

具体实现步骤如下。

1. 使用 STS 创建 Maven 项目并添加依赖的 JAR 包

参见 1.3.3 节和 1.3.4 节，使用 STS 创建一个名为 ch1_3 的 Maven Project，并通过 pom.xml 文件添加项目所依赖的 JAR 包。这里只需要添加 spring-context 依赖。pom.xml 文件内容不再赘述。

2. 创建 DAO 层

在 ch1_3 应用的 src/main/java 目录中创建名为 annotation.dao 的包，在该包中创建 TestDao 接口和 TestDaoImpl 实现类，并将实现类 TestDaoImpl 使用@Repository 注解标注为数据访问层。

TestDao 的代码如下：

```
package annotation.dao;
public interface TestDao {
    public void save();
}
```

TestDaoImpl 的代码如下：

```
package annotation.dao;
import org.springframework.stereotype.Repository;
@Repository("testDaoImpl")
/**相当于@Repository,但如果在 service 层使用@Resource(name="testDaoImpl")注入
Bean,testDaoImpl 不能省略。**/
public class TestDaoImpl implements TestDao{
    @Override
    public void save() {
        System.out.println("testDao save");
    }
}
```

3. 创建 Service 层

在 ch1_3 应用的 src/main/java 目录中创建名为 annotation.service 的包,并在该包中创建 TestService 接口和 TestSeviceImpl 实现类,并将实现类 TestSeviceImpl 使用 @Service 注解标注为业务逻辑层。

TestService 的代码如下:

```
package annotation.service;
public interface TestService {
    public void save();
}
```

TestSeviceImpl 的代码如下:

```
package annotation.service;
import javax.annotation.Resource;
import org.springframework.stereotype.Service;
import annotation.dao.TestDao;
@Service("testServiceImpl")//相当于@Service
public class TestSeviceImpl implements TestService{
    @Resource(name="testDaoImpl")
    /**相当于@Autowired,@Autowired 默认按照 Bean 类型注入**/
    private TestDao testDao;
    @Override
    public void save() {
        testDao.save();
        System.out.println("testService save");
    }
}
```

4. 创建 Controller 层

在 ch1_3 应用的 src/main/java 目录中创建名为 annotation.controller 的包,并在该包中创建 TestController 类,并将 TestController 类使用 @Controller 注解标注为控制器层。

TestController 的代码如下:

```
package annotation.controller;
import org.springframework.beans.factory.annotation.Autowired;
import org.springframework.stereotype.Controller;
import annotation.service.TestService;
@Controller
public class TestController {
    @Autowired
    private TestService testService;
```

```
    public void save() {
        testService.save();
        System.out.println("testController save");
    }
}
```

5. 创建配置类

本书尽量不使用 Spring 的 XML 配置文件，而使用注解和 Java 配置。因此，在此需要使用@Configuration 注解创建一个 Java 配置类（相当于一个 Spring 的 XML 配置文件），并通过@ComponentScan 注解扫描使用注解的包（相当于在 Spring 的 XML 配置文件中使用<context:component-scan base-package="Bean 所在的包路径"/>语句）。

在 ch1_3 应用的 src/main/java 目录中创建名为 annotation.config 的包，并在该包中创建名为 ConfigAnnotation 的配置类。

ConfigAnnotation 的代码如下：

```
package annotation.config;
import org.springframework.context.annotation.ComponentScan;
import org.springframework.context.annotation.Configuration;
@Configuration//声明当前类是一个配置类(见 1.4.4 节)，相当于一个 Spring 的 XML 配置
文件。
@ComponentScan("annotation")
//自动扫描 annotation 包及其子包下使用的注解，并注册为 Bean。
/*相当于在 Spring 的 XML 配置文件中使用<context:component-scan base-package=
"Bean所在的包路径"/>语句。*/
public class ConfigAnnotation {
}
```

6. 创建测试类

在 ch1_3 应用的 src/main/java 目录中创建名为 annotation.test 的包，并在该包中创建测试类 TestAnnotation，具体代码如下：

```
package annotation.test;
import org.springframework.context.annotation.AnnotationConfigApplicationContext;
import annotation.config.ConfigAnnotation;
import annotation.controller.TestController;
public class TestAnnotation {
    public static void main(String[] args) {
        //初始化 Spring 容器 ApplicationContext
        AnnotationConfigApplicationContext appCon =
        new AnnotationConfigApplicationContext(ConfigAnnotation.class);
        TestController tc =appCon.getBean(TestController.class);
        tc.save();
```

```
        appCon.close();
    }
}
```

7. 运行结果

运行测试类 TestAnnotation 的 main 方法，结果如图 1.24 所示。

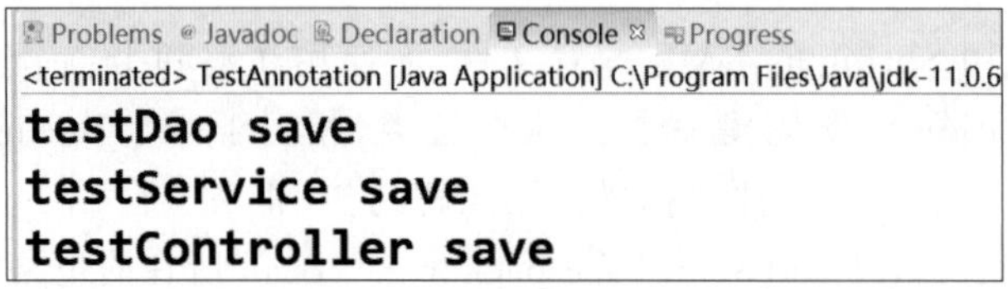

```
testDao save
testService save
testController save
```

图 1.24 ch1_3 的运行结果

1.4.4　Java 配置

Java 配置是 Spring 4.x 推荐的配置方式，是通过@Configuration 和@Bean 注解来实现的。@Configuration 注解声明当前类是一个配置类，相当于一个 Spring 配置的 XML 文件。@Bean 注解在方法上，声明当前方法的返回值为一个 Bean。下面通过实例演示 Java 配置的使用过程。

【例 1-4】　Java 配置的使用过程。该实例的具体要求是：在 DAO 层、Service 层以及 Controller 层中不使用对应注解标注，而是在 Java 配置类中使用@Bean 注解定义它们。

具体实现步骤如下。

1. 使用 STS 创建 Maven 项目并添加依赖的 JAR 包

使用 STS 创建一个名为 ch1_4 的 Maven Project，并通过 pom.xml 文件添加项目依赖的 JAR 包。这里只需要添加 spring-context 依赖，pom.xml 文件内容不再赘述。

2. 创建 DAO 层

在 ch1_4 应用的 src/main/java 目录中创建名为 dao 的包，并在该包中创建 TestDao 类。此类中没有使用@Repository 注解标注为数据访问层，具体代码如下：

```
package dao;
//此处没有使用@Repository 声明 Bean
public class TestDao {
    public void save() {
        System.out.println("TestDao save");
    }
}
```

3. 创建 Service 层

在 ch1_4 应用的 src/main/java 目录中创建名为 service 的包,并在该包中创建 TestService 类。此类中没有使用@Service 注解标注为业务逻辑层,具体代码如下:

```
package service;
import dao.TestDao;
//此处没有使用@Service 声明 Bean
public class TestService {
    //此处没有使用@Autowired 注入 testDao
    TestDao testDao;
    public void setTestDao(TestDao testDao) {
        this.testDao = testDao;
    }
    public void save() {
        testDao.save();
    }
}
```

4. 创建 Controller 层

在 ch1_4 应用的 src/main/java 目录中创建名为 controller 的包,并在该包中创建 TestController 类。此类中没有使用@Controller 注解标注为控制器层,具体代码如下:

```
package controller;
import service.TestService;
//此处没有使用@Controller 声明 Bean
public class TestController {
    //此处没有使用@Autowired 注入 testService
    TestService testService;
    public void setTestService(TestService testService) {
        this.testService = testService;
    }
    public void save() {
        testService.save();
    }
}
```

5. 创建配置类

在 ch1_4 应用的 src/main/java 目录中创建名为 javaConfig 的包,并在该包中创建名为 JavaConfig 的配置类。此类中使用@Configuration 注解标注该类为一个配置类,相当于一个 Spring 配置的 XML 文件。在配置类中使用@Bean 注解定义 0 个或多个 Bean,具体代码如下:

```java
package javaConfig;
import org.springframework.context.annotation.Bean;
import org.springframework.context.annotation.Configuration;
import controller.TestController;
import dao.TestDao;
import service.TestService;
@Configuration
//一个配置类,相当于一个Spring配置的XML文件;
//此处没有使用包扫描,是因为所有Bean都在此类中通过@Bean注解定义了。
public class JavaConfig {
    @Bean
    public TestDao getTestDao() {
        return new TestDao();
    }
    @Bean
    public TestService getTestService() {
        TestService ts = new TestService();
        //使用set方法注入testDao
        ts.setTestDao(getTestDao());
        return ts;
    }
    @Bean
    public TestController getTestController() {
        TestController tc = new TestController();
        //使用set方法注入testService
        tc.setTestService(getTestService());
        return tc;
    }
}
```

6. 创建测试类

在ch1_4应用的javaConfig包中创建测试类TestConfig,具体代码如下:

```java
package javaConfig;
import org.springframework.context.annotation.AnnotationConfigApplicationContext;
import controller.TestController;
public class TestConfig {
    public static void main(String[] args) {
        //初始化Spring容器ApplicationContext
        AnnotationConfigApplicationContext appCon =
            new AnnotationConfigApplicationContext(JavaConfig.class);
        TestController tc = appCon.getBean(TestController.class);
        tc.save();
        appCon.close();
```

            }
        }

7. 运行结果

运行测试类 TestConfig 的 main 方法，运行结果如图 1.25 所示。

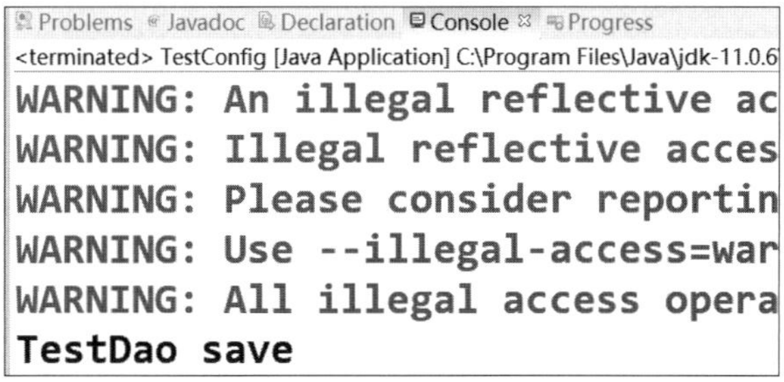

图 1.25　ch1_4 的运行结果

从应用 ch1_3 与应用 ch1_4 对比可以看出，有时使用 Java 配置反而更加烦琐。何时使用 Java 配置？何时使用注解配置？作者的观点是：全局配置尽量使用 Java 配置，如数据库相关的配置；业务 Bean 的配置尽量使用注解配置，如数据访问层、业务逻辑层、控制器层等相关的配置。

1.4.5　实践环节

参考【例 1-4】，将【例 1-2】中的 applicationContext.xml 配置文件替换为 Java 配置类。

1.5　Spring AOP

Spring AOP 是 Spring 框架体系结构中非常重要的功能模块之一，该模块提供了面向切面编程实现。面向切面编程在事务处理、日志记录、安全控制等操作中被广泛使用。

1.5.1　Spring AOP 的基本概念

1. AOP 的概念

AOP，即面向切面编程。它与面向对象编程（Object-Oriented Programming，OOP）相辅相成，提供了与 OOP 不同的抽象软件结构的视角。OOP 以类作为程序的基本单元，而 AOP 中的基本单元是 Aspect（切面）。Struts2 的拦截器设计就是基于 AOP 的思想，是个比较经典的应用。

业务处理代码通常都有日志记录、性能统计、安全控制、事务处理、异常处理等操作。尽管使用 OOP 可以通过封装或继承的方式达到代码重用的功能，但仍然存在同样的代码分散到各个方法中。因此，采用 OOP 处理日志记录等操作不仅增加了开发者的工作量，而且提高了升级维护的困难。为了解决此类问题，AOP 思想应运而生。AOP 采取横向抽取机制,即将分散在各个方法中的重复代码提取出来,然后在程序编译或运行阶段再将这些抽取出来的代码应用到需要执行的地方。这种横向抽取机制无法采用传统的OOP办到,因为OOP实现的是父子关系的纵向重用。但是AOP不是OOP的替代品,而是OOP的补充,它们相辅相成。

在AOP中,横向抽取机制的类与切面的关系如图1.26所示。

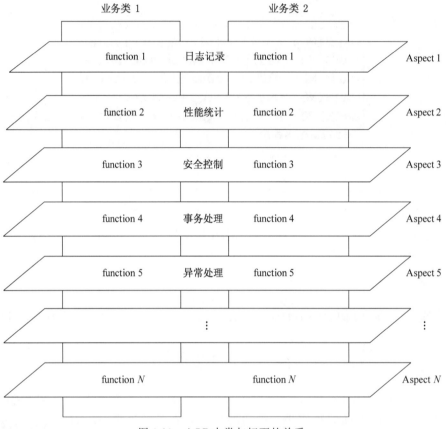

图1.26　AOP中类与切面的关系

从图1.26可以看出,通过切面Aspect分别在业务类1和业务类2中加入了日志记录、性能统计、安全控制、事务处理、异常处理等操作。

2. AOP的术语

Spring AOP框架涉及以下常用术语。

(1) 切面。

切面(aspect)是指封装横切到系统功能(如事务处理)的类。

(2) 连接点。

连接点(joinpoint)是指程序运行中的一些时间点,如方法的调用或异常地抛出。

(3) 切入点。

切入点(pointcut)是指那些需要处理的连接点。在Spring AOP中,所有的方法执行都是连接点,而切入点是一个描述信息,它修饰的是连接点,通过切入点确定哪些连接点需要被处理。切面、连接点和切入点的关系如图 1.27 所示。

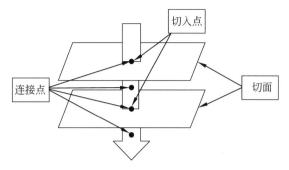

图 1.27　切面、连接点和切入点的关系

(4) 通知(增强处理)。

由切面添加到特定的连接点(满足切入点规则)的一段代码,即在定义好的切入点处所要执行的程序代码。可以将其理解为切面开启后切面的方法。因此,通知是切面的具体实现。

(5) 引入。

引入(introduction)允许在现有的实现类中添加自定义的方法和属性。

(6) 目标对象。

目标对象(target object)是指所有被通知的对象。如果 AOP 框架使用运行时代理的方式(动态的 AOP)来实现切面,那么通知对象总是一个代理对象。

(7) 代理。

代理(proxy)是通知应用到目标对象之后被动态创建的对象。

(8) 组入。

组入(weaving)是将切面代码插入目标对象,从而生成代理对象的过程。根据不同的实现技术,AOP 组入有三种方式:编译器组入,需要特殊的 Java 编译器;类装载期组入,需要特殊的类装载器;动态代理组入,在运行期为目标类添加通知生成子类的方式。Spring AOP 框架默认采用动态代理组入,而 AspectJ(基于 Java 语言的 AOP 框架)采用编译器组入和类装载器组入。

1.5.2　基于注解开发 AspectJ

基于注解开发 AspectJ 要比基于 XML 配置开发 AspectJ 便捷许多,所以在实际开发中推荐使用注解方式。讲解 AspectJ 之前,先讲解 Spring 的通知类型。根据 Spring 中通知在目标类方法的连接点位置,通知可以分为如下 6 种类型。

1. 环绕通知

环绕通知是在目标方法执行前和执行后实施增强,可以应用于日志记录、事务处理等。

2. 前置通知

前置通知是在目标方法执行前实施增强,可应用于权限管理等。

3. 后置返回通知

后置返回通知是在目标方法成功执行后实施增强,可应用于关闭流、删除临时文件等。

4. 后置(最终)通知

后置通知是在目标方法执行后实施增强,与后置返回通知不同的是,不管是否发生异常,都要执行该通知,可应用于释放资源。

5. 异常通知

异常通知是在方法抛出异常后实施增强,可以应用于处理异常、记录日志等。

6. 引入通知

引入通知是在目标类中添加一些新的方法和属性,可以应用于修改目标类(增强类)。

有关 AspectJ 的注解,如表 1.1 所示。

表 1.1 AspectJ 注解

注解名称	描 述
@Aspect	用于定义一个切面,注解在切面类上
@Pointcut	用于定义切入点表达式。使用时,需要定义一个切入点方法。该方法是一个返回值 void,且方法体为空的普通方法
@Before	用于定义前置通知。使用时,通常为其指定 value 属性值,该值既可以是已有的切入点,也可以直接定义切入点表达式
@AfterReturning	用于定义后置返回通知。使用时,通常为其指定 value 属性值,该值既可以是已有的切入点,也可以直接定义切入点表达式
@Around	用于定义环绕通知。使用时,通常为其指定 value 属性值,该值既可以是已有的切入点,也可以直接定义切入点表达式
@AfterThrowing	用于定义异常通知。使用时,通常为其指定 value 属性值,该值既可以是已有的切入点,也可以直接定义切入点表达式。另外,还有一个 throwing 属性用于访问目标方法抛出的异常,该属性值与异常通知方法中同名的形参一致
@After	用于定义后置(最终)通知。使用时,通常为其指定 value 属性值,该值既可以是已有的切入点,也可以直接定义切入点表达式

下面通过一个实例讲解基于注解开发 AspectJ 的过程。

【例 1-5】 基于注解开发 AspectJ 的过程。

该实例的具体要求是:首先,在 DAO 层的实现类中定义 save、modify 和 delete 三个待增强的方法;然后使用@Aspect 注解定义一个切面,在该切面中定义各类型通知,增强

DAO 层中的 save、modify 和 delete 方法。

具体实现步骤如下。

1. 使用 STS 创建 Maven 项目并添加依赖的 JAR 包

使用 STS 创建一个名为 ch1_5 的 Maven Project，并通过 pom.xml 文件添加项目依赖的 JAR 包。spring-aspects 是 Spring 为 AspectJ 提供的实现，因此，ch1_5 除了添加 spring-context 依赖，还需要添加 spring-aspects 依赖。ch1_5 的 pom.xml 文件内容如下：

```xml
<project xmlns="http://maven.apache.org/POM/4.0.0"
    xmlns:xsi="http://www.w3.org/2001/XMLSchema-instance"
    xsi:schemaLocation="http://maven.apache.org/POM/4.0.0
https://maven.apache.org/xsd/maven-4.0.0.xsd">
    <modelVersion>4.0.0</modelVersion>
    <groupId>com.cn.chenheng</groupId>
    <artifactId>ch1_5</artifactId>
    <version>0.0.1-SNAPSHOT</version>
    <packaging>war</packaging>
    <properties>
        <!--spring 版本号 -->
        <spring.version>5.2.3.RELEASE</spring.version>
    </properties>
    <dependencies>
        <dependency>
            <groupId>org.springframework</groupId>
            <artifactId>spring-context</artifactId>
            <version>${spring.version}</version>
        </dependency>
        <dependency>
            <groupId>org.springframework</groupId>
            <artifactId>spring-aspects</artifactId>
            <version>${spring.version}</version>
        </dependency>
    </dependencies>
</project>
```

2. 创建接口及实现类

在 ch1_5 的 src/main/java 目录下创建一个名为 aspectj.dao 的包，并在该包中创建接口 TestDao 和接口实现类 TestDaoImpl。该实现类作为目标类，在切面类中对其方法进行增强处理。使用注解 @Repository，将目标类 aspectj.dao.TestDaoImpl 注解为目标对象。

TestDao 的代码如下：

```java
package aspectj.dao;
public interface TestDao {
    public void save();
    public void modify();
    public void delete();
}
```

TestDaoImpl 的代码如下：

```java
package aspectj.dao;
import org.springframework.stereotype.Repository;
@Repository("testDao")
public class TestDaoImpl implements TestDao{
    @Override
    public void save() {
        System.out.println("保存");
    }
    @Override
    public void modify() {
        System.out.println("修改");
    }
    @Override
    public void delete() {
        System.out.println("删除");
    }
}
```

3. 创建切面类

在 ch1_5 的 src/main/java 目录下创建一个名为 aspectj.annotation 的包，并在该包中创建切面类 MyAspect。在该类中，首先使用@Aspect 注解定义一个切面类，由于该类在 Spring 中是作为组件使用的，所以还需要使用@Component 注解。然后，使用@Pointcut 注解定义切入点表达式，并通过定义方法来表示切入点名称。最后在每个通知方法上添加相应的注解，并将切入点名称作为参数传递给需要执行增强的通知方法。

MyAspect 的代码如下：

```java
package aspectj.annotation;
import org.aspectj.lang.JoinPoint;
import org.aspectj.lang.ProceedingJoinPoint;
import org.aspectj.lang.annotation.After;
import org.aspectj.lang.annotation.AfterReturning;
import org.aspectj.lang.annotation.AfterThrowing;
import org.aspectj.lang.annotation.Around;
import org.aspectj.lang.annotation.Aspect;
import org.aspectj.lang.annotation.Before;
```

```java
import org.aspectj.lang.annotation.Pointcut;
import org.springframework.stereotype.Component;
/**
 * 切面类,在此类中编写各种类型通知
 */
@Aspect//@Aspect 声明一个切面
@Component//@Component 让此切面成为 Spring 容器管理的 Bean
public class MyAspect {
    /**
     * 定义切入点,通知增强哪些方法。
"execution( * aspectj.dao.*.*(..))" 是定义切入点表达式,
该切入点表达式的意思是匹配 aspectj.dao 包中任意类的任意方法的执行。
其中 execution()是表达式的主体,第一个 * 表示的是返回类型,使用 * 代表所有类型;
aspectj.dao 表示的是需要匹配的包名,后面第二个 * 表示的是类名,使用 * 代表匹配包中所有
的类;第三个 * 表示的是方法名,使用 * 表示所有方法;后面(..)表示方法的参数,其中".."表示
任意参数。另外,注意第一个 * 与包名之间有一个空格。
     */
    @Pointcut("execution( * aspectj.dao.*.*(..))")
    private void myPointCut() {
    }
    /**
     * 前置通知,使用 Joinpoint 接口作为参数获得目标对象信息
     */
    @Before("myPointCut()")//myPointCut()是切入点的定义方法
    public void before(JoinPoint jp) {
        System.out.print("前置通知:模拟权限控制");
        System.out.println(",目标类对象:" +jp.getTarget()
        +",被增强处理的方法:" +jp.getSignature().getName());
    }
    /**
     * 后置返回通知
     */
    @AfterReturning("myPointCut()")
    public void afterReturning(JoinPoint jp) {
        System.out.print("后置返回通知:" +"模拟删除临时文件");
        System.out.println(",被增强处理的方法:" +jp.getSignature().getName());
    }
    /**
     * 环绕通知
     * ProceedingJoinPoint 是 JoinPoint 子接口,代表可以执行的目标方法
     * 返回值类型必须是 Object
     * 必须一个参数是 ProceedingJoinPoint 类型
     * 必须 throws Throwable
     */
```

```java
    @Around("myPointCut()")
    public Object around(ProceedingJoinPoint pjp) throws Throwable{
        //开始
        System.out.println("环绕开始:执行目标方法前,模拟开启事务");
        //执行当前目标方法
        Object obj =pjp.proceed();
        //结束
        System.out.println("环绕结束:执行目标方法后,模拟关闭事务");
        return obj;
    }
    /**
     * 异常通知
     */
    @AfterThrowing(value="myPointCut()",throwing="e")
    public void except(Throwable e) {
        System.out.println("异常通知:" +"程序执行异常" +e.getMessage());
    }
    /**
     * 后置(最终)通知
     */
    @After("myPointCut()")
    public void after() {
        System.out.println("最终通知:模拟释放资源");
    }
}
```

4. 创建配置类

在 ch1_5 的 src/main/java 目录下创建一个名为 aspectj.config 的包,并在该包中创建配置类 AspectjAOPConfig。在该类中使用@Configuration 注解声明此类为配置类;使用 @ComponentScan("aspectj") 注解自动扫描 aspectj 包下使用的注解;使用 @EnableAspectJAutoProxy 注解开启 Spring 对 AspectJ 的支持。

AspectjAOPConfig 的代码如下：

```java
package aspectj.config;
import org.springframework.context.annotation.ComponentScan;
import org.springframework.context.annotation.Configuration;
import org.springframework.context.annotation.EnableAspectJAutoProxy;
@Configuration//声明一个配置类
@ComponentScan("aspectj")//自动扫描 aspectj 包下使用的注解
@EnableAspectJAutoProxy//开启 Spring 对 AspectJ 的支持
public class AspectjAOPConfig {
}
```

5. 创建测试类

在 ch1_5 应用的 aspectj.config 包中创建测试类 AOPTest。
AOPTest 的代码如下：

```
package aspectj.config;
import org.springframework.context.annotation.AnnotationConfigApplicationContext;
import aspectj.dao.TestDao;
public class AOPTest {
    public static void main(String[] args) {
        //初始化 Spring 容器 ApplicationContext
        AnnotationConfigApplicationContext appCon =
            new AnnotationConfigApplicationContext(AspectjAOPConfig.class);
        //从容器中获取增强后的目标对象
        TestDao testDaoAdvice = appCon.getBean(TestDao.class);
        //执行方法
        testDaoAdvice.save();
        System.out.println("================");
        testDaoAdvice.modify();
        System.out.println("================");
        testDaoAdvice.delete();
        appCon.close();
    }
}
```

6. 运行测试类

运行测试类 AOPTest 的 main 方法，运行结果如图 1.28 所示。

图 1.28 ch1_5 应用的运行结果

1.6 Spring Bean

在 Spring 的应用中，Spring IoC 容器可以创建、装配和配置应用组件对象，这里的组件对象称为 Bean。

1.6.1 Bean 的实例化

在面向对象编程中，想使用某个对象时，需要事先实例化该对象。同样，在 Spring 框架中，想使用 Spring 容器中的 Bean，也需要实例化 Bean。Spring 框架实例化 Bean 有三种方式：构造方法实例化、静态工厂实例化和实例工厂实例化（其中，最常用的实例方法是构造方法实例化）。

下面通过一个实例 ch1_6 演示 Bean 的实例化过程。

【例 1-6】 Bean 的实例化过程。

该实例的具体要求是：分别使用构造方法、静态工厂和实例工厂实例化 Bean。

具体实现步骤如下。

1. 使用 STS 创建 Maven 项目并添加依赖的 JAR 包

使用 STS 创建一个名为 ch1_6 的 Maven Project，这里只需要添加 spring-context 依赖，pom.xml 文件内容不再赘述。

2. 创建实例化 Bean 的类

在 ch1_6 应用的 src/main/java 目录下创建一个名为 instance 的包，并在该包中创建 BeanClass、BeanInstanceFactory 以及 BeanStaticFactory 等实例化 Bean 的类。

BeanClass 的代码如下：

```
package instance;
public class BeanClass {
    public String message;
    public BeanClass() {
        message = "构造方法实例化 Bean";
    }
    public BeanClass(String s) {
        message = s;
    }
}
```

BeanInstanceFactory 的代码如下：

```
package instance;
public class BeanInstanceFactory {
    public BeanClass createBeanClassInstance() {
```

```
        return new BeanClass("调用实例工厂方法实例化 Bean");
    }
}
```

BeanStaticFactory 的代码如下：

```
package instance;
public class BeanStaticFactory {
    private static BeanClass beanInstance = new BeanClass("调用静态工厂方法实例化 Bean");
    public static BeanClass createInstance() {
        return beanInstance;
    }
}
```

3. 创建配置类

在 ch1_6 应用的 src/main/java 目录下创建一个名为 config 的包，并在该包中创建配置类 JavaConfig。在该配置类中使用@Bean 定义 3 个 Bean，具体代码如下：

```
package config;
import org.springframework.context.annotation.Bean;
import org.springframework.context.annotation.Configuration;
import instance.BeanClass;
import instance.BeanInstanceFactory;
import instance.BeanStaticFactory;
@Configuration
public class JavaConfig {
    /**
     * 构造方法实例化
     */
    @Bean(value="beanClass")//value 可以省略
    public BeanClass getBeanClass() {
        return new BeanClass();
    }
    /**
     * 静态工厂实例化
     */
    @Bean(value="beanStaticFactory")
    public BeanClass getBeanStaticFactory() {
        return BeanStaticFactory.createInstance();
    }
    /**
     * 实例工厂实例化
     */
```

```
    @Bean(value="beanInstanceFactory")
    public BeanClass getBeanInstanceFactory() {
        BeanInstanceFactory bi =new BeanInstanceFactory();
        return bi.createBeanClassInstance();
    }
}
```

4．创建测试类

在 ch1_6 应用的 config 包中创建测试类 TestBean，在该类中测试配置类定义的 Bean，具体代码如下：

```
package config;
import org.springframework.context.annotation.AnnotationConfigApplicationContext;
import instance.BeanClass;
public class TestBean {
    public static void main(String[] args) {
        //初始化 Spring 容器 ApplicationContext
        AnnotationConfigApplicationContext appCon =
            new AnnotationConfigApplicationContext(JavaConfig.class);
        BeanClass b1 =(BeanClass)appCon.getBean("beanClass");
        System.out.println(b1+b1.message);
        BeanClass b2 =(BeanClass)appCon.getBean("beanStaticFactory");
        System.out.println(b2+b2.message);
        BeanClass b3 =(BeanClass)appCon.getBean("beanInstanceFactory");
        System.out.println(b3+b3.message);
        appCon.close();
    }
}
```

5．运行测试类

运行测试类 TestBean 的 main 方法，运行结果如图 1.29 所示。

```
instance.BeanClass@593aaf41构造方法实例化Bean
instance.BeanClass@5a56cdac调用静态工厂方法实例化Bean
instance.BeanClass@7c711375调用实例工厂方法实例化Bean
```

图 1.29　ch1_6 应用的运行结果

1.6.2　Bean 的作用域

在 Spring 中，不仅可以完成 Bean 的实例化，还可以为 Bean 指定作用域。Spring 为 Bean 的实例定义了表 1.2 所示的作用域，通过@Scope 注解实现。

表 1.2　Bean 的作用域

作用域名称	描述
singleton	默认的作用域，使用 singleton 定义的 Bean，Spring 容器中只有一个 Bean 实例
prototype	Spring 容器每次获取 prototype 定义的 Bean，容器都将创建一个新的 Bean 实例
request	在一次 HTTP 请求中，容器将返回一个 Bean 实例，不同的 HTTP 请求返回不同的 Bean 实例。仅在 Web Spring 应用程序上下文中使用
session	在一个 HTTP Session 中，容器将返回同一个 Bean 实例。仅在 Web Spring 应用程序上下文中使用
application	为每个 ServletContext 对象创建一个实例，即同一个应用共享一个 Bean 实例。仅在 Web Spring 应用程序上下文中使用
websocket	为每个 WebSocket 对象创建一个 Bean 实例。仅在 Web Spring 应用程序上下文中使用

在表 1.2 所示的 6 种作用域中，singleton 和 prototype 是最常用的两种，后面 4 种作用域仅使用在 Web Spring 应用程序上下文中。下面通过一个实例演示 Bean 的作用域。

【例 1-7】 Bean 的作用域。

该实例的具体要求是：分别定义作用域为 singleton 和 prototype 的两个 Bean。

具体实现步骤如下。

1. 使用 STS 创建 Maven 项目并添加依赖的 JAR 包

使用 STS 创建一个名为 ch1_7 的 Maven Project，这里只需要添加 spring-context 依赖，pom.xml 文件内容不再赘述。

2. 编写不同作用域的 Bean

在 ch1_7 应用的 src/main/java 目录下创建一个名为 service 的包，并在该包中创建 SingletonService 和 PrototypeService 类。在 SingletonService 类中，Bean 的作用域为默认作用域 singleton；在 PrototypeService 类中，Bean 的作用域为 prototype。

SingletonService 的代码如下：

```
package service;
import org.springframework.stereotype.Service;
@Service//默认为 singleton 相当于@Scope("singleton")
public class SingletonService {
}
```

PrototypeService 的代码如下：

```
package service;
import org.springframework.context.annotation.Scope;
import org.springframework.stereotype.Service;
@Service
```

```
@Scope("prototype")
public class PrototypeService {
}
```

3. 创建配置类

在 ch1_7 应用的 src/main/java 目录下创建一个名为 config 的包，并在该包中创建配置类 ScopeConfig，具体代码如下：

```
package config;
import org.springframework.context.annotation.ComponentScan;
import org.springframework.context.annotation.Configuration;
@Configuration
@ComponentScan("service")
public class ScopeConfig {
}
```

4. 创建测试类

在 ch1_7 应用的 config 包中创建测试类 TestScope，在该测试类中分别获得 SingletonService 和 PrototypeService 的两个 Bean 实例，具体代码如下：

```
package config;
import org.springframework.context.annotation.AnnotationConfigApplicationContext;
import service.PrototypeService;
import service.SingletonService;
public class TestScope {
    public static void main(String[] args) {
        //初始化 Spring 容器 ApplicationContext
        AnnotationConfigApplicationContext appCon =
            new AnnotationConfigApplicationContext(ScopeConfig.class);
        SingletonService ss1 =appCon.getBean(SingletonService.class);
        SingletonService ss2 =appCon.getBean(SingletonService.class);
        System.out.println(ss1);
        System.out.println(ss2);
        PrototypeService ps1 =appCon.getBean(PrototypeService.class);
        PrototypeService ps2 =appCon.getBean(PrototypeService.class);
        System.out.println(ps1);
        System.out.println(ps2);
        appCon.close();
    }
}
```

5. 运行测试类

运行测试类 TestScope 的 main 方法，运行结果如图 1.30 所示。

```
Problems  Javadoc  Declaration  Console  Progress
<terminated> TestScope [Java Application] C:\Program Files\Java\jdk-
service.SingletonService@53aac487
service.SingletonService@53aac487
service.PrototypeService@52b1beb6
service.PrototypeService@273e7444
```

图 1.30 ch1_7 应用的运行结果

从图 1.30 运行结果可以得知,两次获取 SingletonService 的 Bean 实例时,IoC 容器返回两个相同的 Bean 实例;而两次获取 PrototypeService 的 Bean 实例时,IoC 容器返回两个不同的 Bean 实例。

1.6.3 Bean 的初始化和销毁

在实际工程应用中,经常需要在 Bean 使用之前或之后做一些必要的操作,Spring 为 Bean 生命周期的操作提供了支持。可以使用 @Bean 注解的 initMethod 和 destroyMethod 属性(相当于 XML 配置的 init-method 和 destroy-method)对 Bean 进行初始化和销毁。下面通过一个实例演示 Bean 的初始化和销毁。

【例 1-8】 Bean 的初始化和销毁。

该实例的具体要求是:首先,定义一个 MyService 类,在该类中定义构造方法、初始化方法和销毁方法。然后,在 Java 配置类使用 @Bean 注解的 initMethod 和 destroyMethod 属性对 MyService 对象进行初始化和销毁。

具体实现步骤如下。

1. 使用 STS 创建 Maven 项目并添加依赖的 JAR 包

使用 STS 创建一个名为 ch1_8 的 Maven Project,这里只需要添加 spring-context 依赖,pom.xml 文件内容不再赘述。

2. 创建 Bean 的类

在 ch1_8 应用的 src/main/java 目录下创建一个名为 service 的包,并在该包中创建 MyService 类,具体代码如下:

```java
package service;
public class MyService {
    public void initService() {
        System.out.println("initMethod");
    }
    public MyService() {
        System.out.println("构造方法");
    }
    public void destroyService() {
        System.out.println("destroyMethod");
```

```
    }
}
```

3. 创建配置类

在 ch1_8 应用的 src/main/java 目录下创建一个名为 config 的包,并在该包中创建配置类 JavaConfig,具体代码如下:

```
package config;
import org.springframework.context.annotation.Bean;
import org.springframework.context.annotation.Configuration;
import service.MyService;
@Configuration
public class JavaConfig {
    //initMethod 和 destroyMethod 指定 MyService 类的 initService 和 destroyService
方法,在构造方法之后、销毁之前执行
    @Bean(initMethod="initService",destroyMethod="destroyService")
    public MyService getMyService() {
        return new MyService();
    }
}
```

4. 创建测试类

在 ch1_8 应用的 config 包中创建测试类 TestInitAndDestroy,具体代码如下:

```
package config;
import org.springframework.context.annotation.AnnotationConfigApplicationContext;
import service.MyService;
public class TestInitAndDestroy {
    public static void main(String[] args) {
        //初始化 Spring 容器 ApplicationContext
        AnnotationConfigApplicationContext appCon =
    new AnnotationConfigApplicationContext(JavaConfig.class);
        MyService ms =   appCon.getBean(MyService.class);
        appCon.close();
    }
}
```

5. 运行测试类

运行测试类 TestInitAndDestroy 的 main 方法,运行结果如图 1.31 所示。

```
🔲 Problems  @ Javadoc  🔲 Declaration  🖥 Console ⌧
<terminated> TestInitAndDestroy [Java Application]
构造方法
initMethod
destroyMethod
```

图 1.31 ch1_8 应用的运行结果

1.7 Spring 的数据库编程

数据库编程是互联网编程的基础，Spring 框架为开发者提供了 JDBC 模板模式，即 jdbcTemplate，它可以简化许多代码，但在实际应用中并不常用。更多的时候，用的是 Hibernate 框架和 MyBatis 框架进行数据库编程。本节仅简要介绍 Spring jdbcTemplate 的使用方法，而 Hibernate 框架和 MyBatis 框架的相关内容不属于本节内容。

1.7.1 Spring JDBC 的 XML 配置

本节 Spring 数据库编程主要使用 Spring JDBC 模块的 core 和 dataSource 包。core 包是 JDBC 的核心功能包，包括常用的 JdbcTemplate 类；dataSource 包是访问数据源的工具类包。使用 Spring JDBC 操作数据库，需要对其进行配置。XML 配置文件示例代码如下：

```xml
<!--配置数据源 -->
< bean  id =" dataSource "  class =" org. springframework. jdbc. datasource.
DriverManagerDataSource">
        <!--MySQL 数据库驱动 -->
        <property name="driverClassName" value="com.mysql.jdbc.Driver"/>
        <!--连接数据库的 URL -->
        < property name="url" value="jdbc:mysql://localhost:3306/springtest?
characterEncoding=utf8"/>
        <!--连接数据库的用户名 -->
        <property name="username" value="root"/>
        <!--连接数据库的密码 -->
        <property name="password" value="root"/>
</bean>
<!--配置 JDBC 模板 -->
<bean id="jdbcTemplate" class="org.springframework.jdbc.core.JdbcTemplate">
        <property name="dataSource" ref="dataSource"/>
</bean>
```

上述示例代码中，配置 JDBC 模板时，需要将 dataSource 注入 jdbcTemplate，而在数据访问层（如 Dao 类）使用 jdbcTemplate 时，也需要将 jdbcTemplate 注入对应的 Bean 中。代码示例如下：

```
...
@Repository
public class TestDaoImpl implements TestDao{
    @Autowired
    //使用配置文件中的 JDBC 模板
    private JdbcTemplate jdbcTemplate;
    ...
}
```

1.7.2　Spring JDBC 的 Java 配置

与 1.7.1 节 XML 配置文件内容等价的 Java 配置示例如下：

```
package config;
import org.springframework.beans.factory.annotation.Value;
import org.springframework.context.annotation.Bean;
import org.springframework.context.annotation.ComponentScan;
import org.springframework.context.annotation.Configuration;
import org.springframework.context.annotation.PropertySource;
import org.springframework.jdbc.core.JdbcTemplate;
import org.springframework.jdbc.datasource.DriverManagerDataSource;
@Configuration //通过该注解来表明该类是一个 Spring 的配置,相当于一个 xml 文件
@ComponentScan(basePackages = "dao") //配置扫描包
@PropertySource(value={"classpath:jdbc.properties"},ignoreResourceNotFound=true)
//配置多个属性文件时 value={"classpath:jdbc.properties","xx","xxx"}
public class SpringJDBCConfig {
    @Value("${jdbc.url}")//注入属性文件 jdbc.properties 中的 jdbc.url
    private String jdbcUrl;
    @Value("${jdbc.driverClassName}")
    private String jdbcDriverClassName;
    @Value("${jdbc.username}")
    private String jdbcUsername;
    @Value("${jdbc.password}")
    private String jdbcPassword;
    /**
     * 配置数据源
     */
    @Bean
    public DriverManagerDataSource dataSource() {
        DriverManagerDataSource myDataSource =new DriverManagerDataSource();
        // 数据库驱动
        myDataSource.setDriverClassName(jdbcDriverClassName);;
        // 相应驱动的 jdbcUrl
        myDataSource.setUrl(jdbcUrl);
```

```
        // 数据库的用户名
        myDataSource.setUsername(jdbcUsername);
        // 数据库的密码
        myDataSource.setPassword(jdbcUsername);
        return myDataSource;
    }
    /**
     * 配置 JdbcTemplate
     */
    @Bean(value="jdbcTemplate")
    public JdbcTemplate getJdbcTemplate() {
    return new JdbcTemplate(dataSource());
    }
}
```

上述 Java 配置示例中，需要事先在 classpath 目录（如应用的 src/main/java 目录）下创建属性文件，示例代码如下：

```
jdbc.driverClassName=com.mysql.jdbc.Driver
jdbc.url=jdbc:mysql://localhost:3306/springtest?characterEncoding=utf8
jdbc.username=root
jdbc.password=root
```

另外，在数据访问层（如 Dao 类）使用 jdbcTemplate 时，也需要将 jdbcTemplate 注入对应的 Bean 中。代码示例如下：

```
...
@Repository
public class TestDaoImpl implements TestDao{
    @Autowired
    //使用配置文件中的 JDBC 模板
    private JdbcTemplate jdbcTemplate;
    ...
}
```

1.7.3 Spring JdbcTemplate 的常用方法

获取 JDBC 模板后，如何使用它是本节将要讲述的内容。首先，需要了解 JdbcTemplate 类的常用方法。该类的常用方法是 update() 和 query()。

（1） public int update(String sql,Object args[])。

该方法可以对数据表进行增加、修改、删除等操作。使用 args[] 设置 SQL 语句中的参数，并返回更新的行数。示例代码如下：

```
String insertSql ="insert into user values(null,?,?)";
Object param1[] ={"chenheng1", "男"};
```

```
jdbcTemplate.update(sql, param1);
```

（2）public List＜T＞ query（String sql，RowMapper＜T＞ rowMapper，Object args[]）。

该方法可以对数据表进行查询操作。rowMapper 将结果集映射到用户自定义的类中(前提是自定义类中的属性与数据表的字段对应)。示例代码如下：

```
String selectSql ="select * from user";
RowMapper< MyUser > rowMapper = new BeanPropertyRowMapper < MyUser > (MyUser.class);
List<MyUser>list =jdbcTemplate.query(sql, rowMapper, null);
```

下面通过一个实例演示 Spring JDBC 的使用过程。

【例 1-9】 Spring JDBC 的使用过程。

该实例的具体要求是：首先，在 MySQL 数据库中创建数据表 user；然后，使用 Spring JDBC 对数据表 user 进行增删改查。

具体实现步骤如下。

1. 使用 STS 创建 Maven 项目并添加依赖的 JAR 包

使用 STS 创建一个名为 ch1_9 的 Maven Project，并通过 pom.xml 文件添加项目所依赖的 JAR 包。在该例中通过 Spring JDBC 模块访问 MySQL 数据库。所以，这里除了添加 spring-context 依赖外，还需要添加 MySQL 连接依赖和 spring-jdbc 依赖。ch1_9 的 pom.xml 文件内容如下：

```xml
<project xmlns="http://maven.apache.org/POM/4.0.0"
    xmlns:xsi="http://www.w3.org/2001/XMLSchema-instance"
    xsi:schemaLocation="http://maven.apache.org/POM/4.0.0
https://maven.apache.org/xsd/maven-4.0.0.xsd">
    <modelVersion>4.0.0</modelVersion>
    <groupId>com.cn.chenheng</groupId>
    <artifactId>ch1_9</artifactId>
    <version>0.0.1-SNAPSHOT</version>
    <packaging>war</packaging>
    <properties>
        <!--spring 版本号 -->
        <spring.version>5.2.3.RELEASE</spring.version>
    </properties>
    <dependencies>
        <dependency>
            <groupId>org.springframework</groupId>
            <artifactId>spring-context</artifactId>
            <version>${spring.version}</version>
        </dependency>
        <dependency>
```

```xml
            <groupId>org.springframework</groupId>
            <artifactId>spring-jdbc</artifactId>
            <version>${spring.version}</version>
        </dependency>
        <dependency>
            <groupId>mysql</groupId>
            <artifactId>mysql-connector-java</artifactId>
            <version>5.1.45</version>
        </dependency>
    </dependencies>
</project>
```

2. 创建属性文件与配置类

本书使用 MySQL 数据库演示有关数据库访问的内容。因此,需要在 ch1_9 应用的 src/main/resources 目录下创建数据库配置的属性文件 jdbc.properties,具体内容如下:

```
jdbc.driverClassName=com.mysql.jdbc.Driver
jdbc.url=jdbc:mysql://localhost:3306/springtest?characterEncoding=utf8
jdbc.username=root
jdbc.password=root
```

在 ch1_9 应用的 src/main/java 目录下创建一个名为 config 的包,并在该包中创建配置类 SpringJDBCConfig。在该配置类中使用@PropertySource 注解读取属性文件 jdbc.properties,并配置数据源和 JdbcTemplate,具体代码如下:

```java
package config;
import org.springframework.beans.factory.annotation.Value;
import org.springframework.context.annotation.Bean;
import org.springframework.context.annotation.ComponentScan;
import org.springframework.context.annotation.Configuration;
import org.springframework.context.annotation.PropertySource;
import org.springframework.jdbc.core.JdbcTemplate;
import org.springframework.jdbc.datasource.DriverManagerDataSource;
@Configuration //通过该注解来表明该类是一个 Spring 的配置,相当于一个 xml 文件
@ComponentScan(basePackages ={"dao","service"}) //配置扫描包
@PropertySource(value={"classpath:jdbc.properties"},ignoreResourceNotFound
=true)
//配置多个属性文件 value={"classpath:jdbc.properties","xx","xxx"}
public class SpringJDBCConfig {
    @Value("${jdbc.url}")//注入属性文件 jdbc.properties 中的 jdbc.url
    private String jdbcUrl;
    @Value("${jdbc.driverClassName}")
    private String jdbcDriverClassName;
    @Value("${jdbc.username}")
```

```
    private String jdbcUsername;
    @Value("${jdbc.password}")
    private String jdbcPassword;
    /**
     * 配置数据源
     */
    @Bean
    public DriverManagerDataSource dataSource() {
        DriverManagerDataSource myDataSource = new DriverManagerDataSource();
        // 数据库驱动
            myDataSource.setDriverClassName(jdbcDriverClassName);;
        // 相应驱动的 jdbcUrl
            myDataSource.setUrl(jdbcUrl);
        // 数据库的用户名
            myDataSource.setUsername(jdbcUsername);
        // 数据库的密码
            myDataSource.setPassword(jdbcUsername);
        return myDataSource;
    }
    /**
     * 配置 JdbcTemplate
     */
    @Bean(value="jdbcTemplate")
    public JdbcTemplate getJdbcTemplate() {
            return new JdbcTemplate(dataSource());
    }
}
```

3. 创建数据表与实体类

使用 Navicat for MySQL 创建数据库 springtest,并在该数据库中创建数据表 user,数据表 user 的结构如图 1.32 所示。

名	类型	长度	小数点	允许空值(	
▶ uid	int	10	0	☐	🔑1
uname	varchar	20	0	☑	
usex	varchar	10	0	☑	

图 1.32　user 表的结构

在 ch1_9 应用的 src/main/java 目录下创建一个名为 entity 的包,在该包中创建实体类 MyUser,具体代码如下:

```
package entity;
public class MyUser {
    private Integer uid;
```

```
    private String uname;
    private String usex;
    //省略 set 和 get 方法
    public String toString() {
        return "myUser [uid=" +uid +", uname=" +uname +", usex=" +usex +"]";
    }
}
```

4. 创建数据访问层

在 ch1_9 应用的 src/main/java 目录下创建一个名为 dao 的包,在该包中创建数据访问接口 TestDao 和接口实现类 TestDaoImpl。在实现类 TestDaoImpl 中使用@Repository 注解标注此类为数据访问层,并使用@Autowired 注解依赖注入 JdbcTemplate。

TestDao 的代码如下:

```
package dao;
import java.util.List;
import entity.MyUser;
public interface TestDao {
    public int update(String sql, Object[] param);
    public List<MyUser>query(String sql, Object[] param);
}
```

TestDaoImpl 的代码如下:

```
package dao;
import java.util.List;
import org.springframework.beans.factory.annotation.Autowired;
import org.springframework.jdbc.core.BeanPropertyRowMapper;
import org.springframework.jdbc.core.JdbcTemplate;
import org.springframework.jdbc.core.RowMapper;
import org.springframework.stereotype.Repository;
import entity.MyUser;
@Repository
public class TestDaoImpl implements TestDao{
    @Autowired
    //使用配置类中的 JDBC 模板
    private JdbcTemplate jdbcTemplate;
    /**
     * 更新方法,包括添加、修改、删除
     * param 为 sql 中的参数,如通配符?
     */
    @Override
    public int update(String sql, Object[] param) {
        return jdbcTemplate.update(sql, param);
```

```
    }
    /**
     * 查询方法
     * param 为 sql 中的参数,如通配符?
     */
    @Override
    public List<MyUser>query(String sql, Object[] param) {
         RowMapper< MyUser > rowMapper = new BeanPropertyRowMapper< MyUser >(MyUser.class);
         return jdbcTemplate.query(sql, rowMapper);
    }
}
```

5.创建业务逻辑层

在 ch1_9 应用的 src/main/java 目录下创建一个名为 service 的包,在该包中创建接口 TestService 和接口实现类 TestServiceImpl。在实现类 TestServiceImpl 中使用 @Service 注解标注此类为业务逻辑层,并使用@Autowired 注解依赖注入 TestDao。

TestService 的代码如下:

```
package service;
public interface TestService {
    public void testJDBC();
}
```

TestServiceImpl 的代码如下:

```
package service;
import java.util.List;
import org.springframework.beans.factory.annotation.Autowired;
import org.springframework.stereotype.Service;
import dao.TestDao;
import entity.MyUser;
@Service
public class TestServiceImpl implements TestService{
    @Autowired
    public TestDao testDao;
    @Override
    public void testJDBC() {
        String insertSql ="insert into user values(null,?,?)";
        //数组 param 的值与 insertSql 语句中?一一对应
        Object param1[] ={"chenheng1", "男"};
        Object param2[] ={"chenheng2", "女"};
        Object param3[] ={"chenheng3", "男"};
        Object param4[] ={"chenheng4", "女"};
```

```
        //添加用户
        testDao.update(insertSql, param1);
        testDao.update(insertSql, param2);
        testDao.update(insertSql, param3);
        testDao.update(insertSql, param4);
        //查询用户
        String selectSql = "select * from user";
        List<MyUser> list = testDao.query(selectSql, null);
        for(MyUser mu : list) {
            System.out.println(mu);
        }
    }
}
```

6. 创建测试类

在 ch1_9 应用的 config 包中创建测试类 TestJDBC，具体代码如下：

```
package config;
import org.springframework.context.annotation.AnnotationConfigApplicationContext;
import service.TestService;
public class TestJDBC {
    public static void main(String[] args) {
        //初始化 Spring 容器 ApplicationContext
        AnnotationConfigApplicationContext appCon =
            new AnnotationConfigApplicationContext(SpringJDBCConfig.class);
        TestService ts = appCon.getBean(TestService.class);
        ts.testJDBC();
        appCon.close();
    }
}
```

7. 运行测试类

运行测试类 TestJDBC 的 main 方法，运行结果如图 1.33 所示。

```
myUser [uid=45, uname=chenheng1, usex=男]
myUser [uid=46, uname=chenheng2, usex=女]
myUser [uid=47, uname=chenheng3, usex=男]
myUser [uid=48, uname=chenheng4, usex=女]
```

图 1.33 ch1_9 应用的运行结果

1.7.4 基于@Transactional注解的声明式事务管理

Spring的声明式事务管理是通过AOP技术实现的事务管理,其本质是对方法前后进行拦截,然后在目标方法开始之前创建或加入一个事务,执行完目标方法之后根据执行情况提交或回滚事务。

声明式事务管理最大的优点是不需要通过编程的方式管理事务,因而不需要在业务逻辑代码中掺杂事务处理的代码只需相关的事务规则声明,便可将事务规则应用到业务逻辑中。在通常情况下,在开发中使用声明式事务处理,不仅因为其简单,更主要的是因为这样使得纯业务代码不被污染,极大方便了后期的代码维护。

和编程式事务管理相比,声明式事务管理唯一不足的地方是最细粒度只能作用到方法级别,无法做到像编程式事务管理那样可以作用到代码块级别。但即便有这样的需求,也可以通过变通的方法解决,比如,可以将需要进行事务处理的代码块独立为方法等。Spring的声明式事务管理可以通过两种方式来实现:一是基于XML的方式;二是基于@Transactional注解的方式。

@Transactional注解可以作用于接口、接口方法、类以及类方法上。作用于类上时,该类的所有public方法将都具有该类型的事务属性,同时,也可以在方法级别使用该注解来覆盖类级别的定义。虽然@Transactional注解可以作用于接口、接口方法、类以及类方法上,但是Spring小组建议不要在接口或接口方法上使用该注解,因为这只有在使用基于接口的代理时才会生效。可以使用@Transactional注解的属性定制事务行为,具体属性如表1.3所示。

表1.3 @Transactional的属性

属 性	属性值含义	默认值
propagation	propagation定义了事务的生命周期,主要有以下选项: ① Propagation.REQUIRED:需要事务支持的方法A被调用时,没有事务新建一个事务。当在方法A中调用另一个方法B时,方法B将使用相同的事务。如果方法B发生异常,需要数据回滚时,整个事务数据回滚。 ② Propagation.REQUIRES_NEW:对于方法A和B,在方法调用时,无论是否有事务,都开启一个新的事务;方法B有异常不会导致方法A的数据回滚。 ③ Propagation.NESTED:和Propagation.REQUIRES_NEW类似,仅支持JDBC,不支持JPA或Hibernate。 ④ Propagation.SUPPORTS:方法调用时有事务就使用事务,没有事务就不创建事务。 ⑤ Propagation.NOT_SUPPORTED:强制方法在事务中执行,若有事务,在方法调用到结束阶段事务都将会被挂起。 ⑥ Propagation.NEVER:强制方法不在事务中执行,若有事务则抛出异常。 ⑦ Propagation.MANDATORY:强制方法在事务中执行,若无事务则抛出异常	Propagation.REQUIRED

续表

属　　性	属性值含义	默认值
isolation	isolation(隔离)决定了事务的完整性,处理在多事务对相同数据下的处理机制,主要包含以下隔离级别(前提是当前数据库是否支持)： ① Isolation.READ_UNCOMMITTED：对于在 A 事务里修改了一条记录但没有提交事务,在 B 事务中可以读取到修改后的记录。可导致脏读、不可重复读以及幻读。 ② Isolation.READ_COMMITTED：只有当在 A 事务里修改了一条记录且提交事务后,B 事务才可以读取到提交后的记录,防止脏读,但可能导致不可重复读和幻读。 ③ Isolation.REPEATABLE_READ：不仅能实现 Isolation.READ_COMMITTED 的功能,而且还能阻止当 A 事务读取了一条记录,B 事务将不允许修改该条记录;阻止脏读和不可重复读,但可出现幻读。 ④ Isolation.SERIALIZABLE：此级别下事务是顺序执行的,可以避免上述级别的缺陷,但开销较大。 ⑤ Isolation.DEFAULT：使用当前数据库的默认隔离级别。如 Oracle 和 SQL Server 是 READ_COMMITTED；MySQL 是 REPEATABLE_READ	Isolation.DEFAULT
timeout	timeout 指定事务过期时间,默认为当前数据库的事务过期时间	
readOnly	指定当前事务是否是只读事务	false
rollbackFor	指定哪个或哪些异常可以引起事务回滚(Class 对象数组,必须继承自 Throwable)	Throwable 的子类
rollbackForClassName	指定哪个或哪些异常可以引起事务回滚(类名数组,必须继承自 Throwable)	Throwable 的子类
noRollbackFor	指定哪个或哪些异常不可以引起事务回滚(Class 对象数组,必须继承自 Throwable)	Throwable 的子类
noRollbackForClassName	指定哪个或哪些异常不可以引起事务回滚(类名数组,必须继承自 Throwable)	Throwable 的子类

本节通过实例演示基于@Transactional 注解的声明式事务管理。

【例 1-10】 基于@Transactional 注解的声明式事务管理。

【例 1-10】是通过修改【例 1-9】中的代码实现的。

具体步骤如下。

1. 修改配置类

在配置类中,使用@EnableTransactionManagement 注解开启声明式事务的支持。同时为数据源添加事务管理器。修改后的配置类代码如下：

```
package config;
import org.springframework.beans.factory.annotation.Value;
import org.springframework.context.annotation.Bean;
```

```java
import org.springframework.context.annotation.ComponentScan;
import org.springframework.context.annotation.Configuration;
import org.springframework.context.annotation.PropertySource;
import org.springframework.jdbc.core.JdbcTemplate;
import org.springframework.jdbc.datasource.DataSourceTransactionManager;
import org.springframework.jdbc.datasource.DriverManagerDataSource;
import org.springframework.transaction.annotation.EnableTransactionManagement;
@Configuration //通过该注解来表明该类是一个Spring的配置,相当于一个xml文件
@ComponentScan(basePackages ={"dao","service"}) //配置扫描包
@PropertySource(value={"classpath:jdbc.properties"},ignoreResourceNotFound
=true)
@EnableTransactionManagement//开启声明式事务的支持
//配置多个配置文件 value={"classpath:jdbc.properties","xx","xxx"}
public class SpringJDBCConfig {
    @Value("${jdbc.url}")//注入属性文件jdbc.properties中的jdbc.url
    private String jdbcUrl;
    @Value("${jdbc.driverClassName}")
    private String jdbcDriverClassName;
    @Value("${jdbc.username}")
    private String jdbcUsername;
    @Value("${jdbc.password}")
    private String jdbcPassword;
    /* 配置数据源*/
    @Bean
    public DriverManagerDataSource dataSource() {
        DriverManagerDataSource myDataSource =new DriverManagerDataSource();
        // 数据库驱动
        myDataSource.setDriverClassName(jdbcDriverClassName);;
        // 相应驱动的jdbcUrl
        myDataSource.setUrl(jdbcUrl);
        // 数据库的用户名
        myDataSource.setUsername(jdbcUsername);
        // 数据库的密码
        myDataSource.setPassword(jdbcUsername);
        return myDataSource;
    }
    /* 配置JdbcTemplate */
    @Bean(value="jdbcTemplate")
    public JdbcTemplate getJdbcTemplate() {
        return new JdbcTemplate(dataSource());
    }
    /*为数据源添加事务管理器 */
```

```
    @Bean
    public DataSourceTransactionManager transactionManager() {
        DataSourceTransactionManager dt =new DataSourceTransactionManager();
        dt.setDataSource(dataSource());
        return dt;
    }
}
```

2. 修改业务逻辑层

在实际开发中，通常通过 Service 层进行事务管理，因此需要为 Service 层添加 @Transactional 注解。

添加@Transactional 注解后的 TestServiceImpl 类的代码如下：

```
package service;
import java.util.List;
import org.springframework.beans.factory.annotation.Autowired;
import org.springframework.stereotype.Service;
import org.springframework.transaction.annotation.Transactional;
import dao.TestDao;
import entity.MyUser;
@Service
@Transactional
public class TestServiceImpl implements TestService{
    @Autowired
    public TestDao testDao;
    @Override
    public void testJDBC() {
        String insertSql ="insert into user values(null,?,?)";
        //数组 param 的值与 insertSql 语句中?一一对应
        Object param1[] ={"chenheng1", "男"};
        Object param2[] ={"chenheng2", "女"};
        Object param3[] ={"chenheng3", "男"};
        Object param4[] ={"chenheng4", "女"};
        String insertSql1 ="insert into user values(?,?,?)";
        Object param5[] ={1,"chenheng5", "女"};
        Object param6[] ={1,"chenheng6", "女"};
        //添加用户
        testDao.update(insertSql, param1);
        testDao.update(insertSql, param2);
        testDao.update(insertSql, param3);
        testDao.update(insertSql, param4);
        //添加两个 ID 相同的用户，出现唯一性约束异常，使事务回滚。
        testDao.update(insertSql1, param5);
```

```
        testDao.update(insertSql1, param6);
    //查询用户
    String selectSql = "select * from user";
    List<MyUser> list = testDao.query(selectSql, null);
    for(MyUser mu : list) {
        System.out.println(mu);
    }
    }
}
```

1.7.5 在事务处理中捕获异常

　　　　　　声明式事务处理的流程是：①Spring 根据配置完成事务定义，设置事务属性。②执行开发者的代码逻辑。③如果开发者的代码产生异常（如主键重复）并且满足事务回滚的配置条件，则事务回滚；否则，事务提交。④事务资源释放。

现在的问题是，如果开发者在代码逻辑中加入了 try…catch…语句，Spring 还能不能在声明式事务处理中正常得到事务回滚的异常信息？答案是不能。例如，将 1.7.4 节中 TestServiceImpl 实现类的 testJDBC 方法的代码修改如下：

```
@Override
public void testJDBC () {
    String insertSql = "insert into user values(null,?,?)";
    //数组 param 的值与 insertSql 语句中?一一对应
    Object param1[] = {"chenheng1", "男"};
    Object param2[] = {"chenheng2", "女"};
    Object param3[] = {"chenheng3", "男"};
    Object param4[] = {"chenheng4", "女"};
    String insertSql1 = "insert into user values(?,?,?)";
    Object param5[] = {1, "chenheng5", "女"};
    Object param6[] = {1, "chenheng6", "女"};
    try {
        //添加用户
        testDao.update(insertSql, param1);
        testDao.update(insertSql, param2);
        testDao.update(insertSql, param3);
        testDao.update(insertSql, param4);
        //添加两个 ID 相同的用户，出现唯一性约束异常，使事务回滚。
        testDao.update(insertSql1, param5);
        testDao.update(insertSql1, param6);
        //查询用户
        String selectSql = "select * from user";
        List<MyUser> list = testDao.query(selectSql, null);
        for(MyUser mu : list) {
            System.out.println(mu);
```

```
        }
    } catch (Exception e) {
        System.out.println("主键重复,事务回滚。");
    }
}
```

这时再运行测试类,发现主键重复但事务并没有回滚。这是因为在默认情况下,Spring 只在发生未被捕获的 RuntimeExcetpion 时才回滚事务。现在,如何在事务处理中捕获异常呢?具体修改如下:

(1) 修改@Transactional 注解。

需要将 TestServiceImpl 类中的@Transactional 注解修改为:

```
@Transactional(rollbackFor={Exception.class})
//rollbackFor 指定回滚生效的异常类,多个异常类逗号分隔;
//noRollbackFor 指定回滚失效的异常类
```

(2) 在 catch 语句中添加 throw new RuntimeException();语句。

注意:在实际工程应用中,经常只需要在 catch 语句中添加 TransactionAspectSupport.currentTransactionStatus().setRollbackOnly();语句即可。也就是说,不需要在@Transaction 注解中添加 rollbackFor 属性。

1.7.6 实践环节

参考【例 1-9】,在 MySQL 中创建一个 student 表,并使用 Spring JDBC 对 student 表进行增删改查(需考虑事务处理)。

1.8 本章小结

本章讲解了 Spring IoC、AOP、Bean 以及事务管理等基础知识,目的是让读者在学习 Spring MVC 之前对其有个简要了解。

习 题 1

1. Spring 的核心容器由哪些模块组成?
2. 如何找到 Spring 框架的官方 API?
3. 什么是 Spring IoC? 什么是依赖注入?
4. 在 Java 配置类中如何开启 Spring 对 AspectJ 的支持? 又如何开启 Spring 对声明式事务的支持?
5. 什么是 Spring AOP? 它与 OOP 是什么关系?
6. 使用 Maven 管理 Spring 应用时,为什么没有添加 commons-logging 依赖?

第 2 章

Spring MVC 入门

学习目的与要求

本章重点讲解 MVC 的设计思想及 Spring MVC 的工作原理。通过本章的学习，了解 Spring MVC 的工作原理，掌握 Spring MVC 应用的开发步骤。

本章主要内容

- Spring MVC 的工作原理。
- 第一个 Spring MVC 应用。

MVC 思想将一个应用分成三个基本部分：Model（模型）、View（视图）和 Controller（控制器），让这三个部分以最低的耦合协同工作，从而提高应用的可扩展性及可维护性。Spring MVC 是一款优秀的基于 MVC 思想的应用框架，它是 Spring 框架提供的一个实现了 Web MVC 设计模式的轻量级 Web 框架。

2.1 MVC 模式与 Spring MVC 工作原理

2.1.1 MVC 模式

1. MVC 的概念

MVC 是 Model、View 和 Controller 的缩写,分别代表 Web 应用程序中的三种职责:

- 模型——用于存储数据以及处理用户请求的业务逻辑。
- 视图——向控制器提交数据,显示模型中的数据。
- 控制器——根据视图提出的请求,判断将请求和数据交给哪个模型处理,处理后的有关结果交给哪个视图更新显示。

2. 基于 Servlet 的 MVC 模式

基于 Servlet 的 MVC 模式的具体实现如下:

- 模型:一个或多个 JavaBean 对象,用于存储数据(实体模型,由 JavaBean 类创建)和处理业务逻辑(业务模型,由一般的 Java 类创建)。
- 视图:一个或多个 JSP 页面,向控制器提交数据和为模型提供数据显示,JSP 页面主要使用 HTML 标记和 JavaBean 标记来显示数据。
- 控制器:一个或多个 Servlet 对象,根据视图提交的请求控制,即将请求转发给处理业务逻辑的 JavaBean,并将处理结果存放到实体模型 JavaBean 中,输出给视图显示。

基于 Servlet 的 MVC 模式的流程如图 2.1 所示。

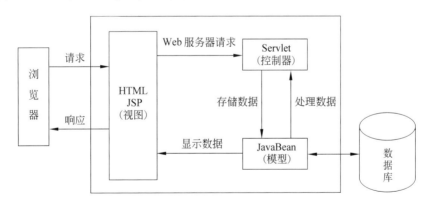

图 2.1　JSP 中的 MVC 模式

2.1.2 Spring MVC 工作原理

Spring MVC 框架是高度可配置的,包含多种视图技术,如 JSP 技术、Velocity、Tiles、

iText 和 POI。Spring MVC 框架并不关心使用的视图技术,也不会强迫开发者只使用 JSP 技术,但本书使用的视图是 JSP。

Spring MVC 框架主要由 DispatcherServlet、处理器映射、控制器、视图解析器、视图组成,其工作原理如图 2.2 所示。

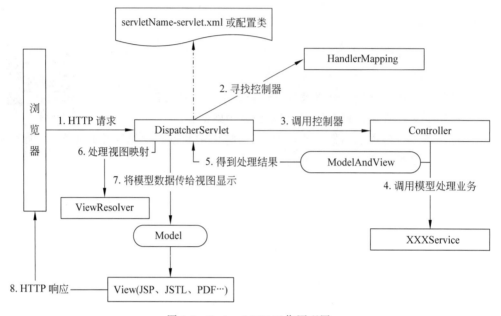

图 2.2 Spring MVC 工作原理图

从图 2.2 可总结出 Spring MVC 的工作流程如下:

(1) 客户端请求提交到 DispatcherServlet。

(2) 由 DispatcherServlet 控制器寻找一个或多个 HandlerMapping,找到处理请求的 Controller。

(3) DispatcherServlet 将请求提交到 Controller。

(4) Controller 调用业务逻辑处理后,返回 ModelAndView。

(5) DispatcherServlet 寻找一个或多个 ViewResoler 视图解析器,找到 ModelAndView 指定的视图。

(6) 视图负责将结果显示到客户端。

2.1.3 Spring MVC 接口

图 2.2 中包含 4 个 Spring MVC 接口:DispatcherServlet、HandlerMapping、Controller 和 ViewResoler。

Spring MVC 所有的请求都经过 DispatcherServlet 统一分发。DispatcherServlet 将请求分发给 Controller 之前,需要借助 Spring MVC 提供的 HandlerMapping 定位到具体的 Controller。

HandlerMapping 接口负责完成客户请求到 Controller 映射。

Controller 接口将处理用户请求,这和 Java Servlet 扮演的角色是一致的。一旦 Controller 处理完用户请求,则返回 ModelAndView 对象给 DispatcherServlet 前端控制器,ModelAndView 中包含了模型(Model)和视图(View)。从宏观考虑,DispatcherServlet 是整个 Web 应用的控制器;从微观考虑,Controller 是单个 HTTP 请求处理过程中的控制器,而 ModelAndView 是 HTTP 请求过程中返回的模型(Model)和视图(View)。

ViewResolver 接口(视图解析器)在 Web 应用中负责查找 View 对象,从而将相应结果渲染给客户。

2.2 第一个 Spring MVC 应用

本节通过一个简单的 Web 应用 ch2_1 来演示 Spring MVC 入门程序的实现过程。

【例 2-1】 Spring MVC 入门程序的实现过程。

该实例的具体要求是:通过应用程序首页面 index.jsp(位于 src/main/webapp 目录)中的超链接打开注册和登录页面(位于 src/main/webapp/WEB-INF/jsp 目录)。

具体实现步骤如下。

2.2.1 创建 Maven 项目并添加依赖的 JAR 包

使用 STS 创建一个名为 ch2_1 的 Maven Project,并通过 pom.xml 文件添加项目所依赖的 JAR 包。该实例演示第一个 Spring MVC 应用。这里应该添加 spring-context、spring-web 和 spring-webmvc 依赖。但是 spring-webmvc 依赖于 spring-context 和 spring-web,所以只需添加 spring-webmvc 依赖即可。ch2_1 的 pom.xml 文件内容如下:

```xml
<project xmlns="http://maven.apache.org/POM/4.0.0"
    xmlns:xsi="http://www.w3.org/2001/XMLSchema-instance"
    xsi:schemaLocation="http://maven.apache.org/POM/4.0.0
https://maven.apache.org/xsd/maven-4.0.0.xsd">
    <modelVersion>4.0.0</modelVersion>
    <groupId>com.cn.chenheng</groupId>
    <artifactId>ch2_1</artifactId>
    <version>0.0.1-SNAPSHOT</version>
    <packaging>war</packaging>
    <properties>
       <!--spring版本号 -->
       <spring.version>5.2.3.RELEASE</spring.version>
    </properties>
    <dependencies>
       <dependency>
          <groupId>org.springframework</groupId>
```

```xml
        <artifactId>spring-webmvc</artifactId>
        <version>${spring.version}</version>
    </dependency>
  </dependencies>
</project>
```

2.2.2 在 web.xml 文件中部署 DispatcherServlet

开发 Spring MVC 应用时,需要在 web.xml 中部署 DispatcherServlet,代码如下:

```xml
<?xml version="1.0" encoding="UTF-8"?>
<web-app id="WebApp_ID" version="4.0"
    xmlns="http://xmlns.jcp.org/xml/ns/javaee"
    xmlns:xsi="http://www.w3.org/2001/XMLSchema-instance"
    xsi:schemaLocation="http://xmlns.jcp.org/xml/ns/javaee
    http://xmlns.jcp.org/xml/ns/javaee/web-app_4_0.xsd">
    <!--部署 DispatcherServlet -->
    <servlet>
        <servlet-name>springmvc</servlet-name>
        <servlet-class>org.springframework.web.servlet.DispatcherServlet
</servlet-class>
        <!--表示容器在启动时立即加载 servlet -->
        <load-on-startup>1</load-on-startup>
    </servlet>
    <servlet-mapping>
        <servlet-name>springmvc</servlet-name>
        <!--处理所有 URL -->
        <url-pattern>/</url-pattern>
    </servlet-mapping>
</web-app>
```

上述 DispatcherServlet 的 servlet 对象 springmvc 初始化时,将在应用程序的 WEB-INF 目录下查找一个配置文件(见 2.2.5 小节),该配置文件的命名规则是 servletName-servlet.xml,如 springmvc-servlet.xml。

另外,也可以将 Spring MVC 的配置文件存放在应用程序的其他地方,但需要使用 servlet 的 init-param 元素加载配置文件。示例代码如下:

```xml
<!--部署 DispatcherServlet-->
<servlet>
    <servlet-name>springmvc</servlet-name>
    <servlet-class>org.springframework.web.servlet.DispatcherServlet
</servlet-class>
    <init-param>
        <param-name>contextConfigLocation</param-name>
        <param-value>/WEN-INF/spring-config/springmvc-servlet.xml
```

```
        </param-value>
    </init-param>
    <load-on-startup>1</load-on-startup>
</servlet>
<servlet-mapping>
    <servlet-name>springmvc</servlet-name>
    <url-pattern>/</url-pattern>
</servlet-mapping>
```

2.2.3 创建 Web 应用首页

在 ch2_1 应用的 src/main/webapp 目录下新建 JSP 文件 index.jsp。index.jsp 的代码如下：

```
<%@page language="java" contentType="text/html; charset=UTF-8"
    pageEncoding="UTF-8"%>
<!DOCTYPE html>
<html>
<head>
<meta charset="UTF-8">
<title>Insert title here</title>
</head>
<body>
    没注册的用户，请<a href="index/register">注册</a>!<br>
    已注册的用户，去<a href="index/login">登录</a>!
</body>
</html>
```

2.2.4 创建 Controller 类

在 ch2_1 应用的 src/main/java 目录下创建一个名为 controller 的包，并在该包中创建基于注解的名为 IndexController 的控制器类，该类中有两个处理请求方法，分别处理首页的"注册"和"登录"超链接请求。

```
package controller;
import org.springframework.stereotype.Controller;
import org.springframework.web.bind.annotation.RequestMapping;
/**"@Controller"表示 IndexController 的实例是一个控制器
 * @Controller 相当于@Controller("indexController")
 * 或@Controller(value = "indexController")
 */
@Controller
@RequestMapping("/index")
public class IndexController {
    @RequestMapping("/login")
```

```
    public String login() {
        /**login 代表逻辑视图名称,需要根据 Spring MVC 配置
         * 文件中 internalResourceViewResolver 的前缀和后缀找到对应的物理视图
         */
        return "login";
    }
    @RequestMapping("/register")
    public String register() {
        return "register";
    }
}
```

2.2.5 创建 Spring MVC 配置文件

在 Spring MVC 中使用扫描机制找到应用中所有基于注解的控制器类。所以,为了让控制器类被 Spring MVC 框架扫描到,需要在配置文件中声明 spring-context,并使用 <context:component-scan/> 元素指定控制器类的基本包(确保所有控制器类都在基本包及其子包下)。另外,需要在配置文件中定义 Spring MVC 的视图解析器(ViewResolver),示例代码如下:

```
<bean class="org.springframework.web.servlet.view.InternalResourceViewResolver"
        id="internalResourceViewResolver">
    <!--前缀 -->
    <property name="prefix" value="/WEB-INF/jsp/" />
    <!--后缀 -->
    <property name="suffix" value=".jsp" />
</bean>
```

上述视图解析器配置了前缀和后缀两个属性。控制器类中视图路径仅需提供 register 和 login,视图解析器将会自动添加前缀和后缀。

在 ch2_1 应用的 src/main/webapp/WEB-INF 目录下创建名为 springmvc-servlet.xml 的配置文件,其代码如下:

```
<?xml version="1.0" encoding="UTF-8"?>
<beans xmlns="http://www.springframework.org/schema/beans"
    xmlns:xsi="http://www.w3.org/2001/XMLSchema-instance"
    xmlns:context="http://www.springframework.org/schema/context"
    xsi:schemaLocation="
    http://www.springframework.org/schema/beans
    http://www.springframework.org/schema/beans/spring-beans.xsd
        http://www.springframework.org/schema/context
        http://www.springframework.org/schema/context/spring-context.xsd">
    <!--使用扫描机制,扫描控制器类 -->
    <context:component-scan base-package="controller"/>
```

```xml
<!--配置视图解析器 -->
<bean class="org.springframework.web.servlet.view.InternalResourceViewResolver"
      id="internalResourceViewResolver">
    <!--前缀 -->
    <property name="prefix" value="/WEB-INF/jsp/" />
    <!--后缀 -->
    <property name="suffix" value=".jsp" />
</bean>
</beans>
```

2.2.6 应用的其他页面

IndexController 控制器的 register 方法处理成功后,跳转到/WEB-INF/jsp/register.jsp 视图;IndexController 控制器的 login 方法处理成功后,跳转到/WEB-INF/jsp/login.jsp 视图。因此,应用的/WEB-INF/jsp 目录下应有 register.jsp 和 login.jsp 页面,这两个 JSP 页面代码略。

2.2.7 发布并运行 Spring MVC 应用

在 STS 中第一次运行 Spring MVC 应用时,需要将应用发布到 Tomcat。例如,运行 ch2_1 应用时,可以选中应用名称 ch2_1,右击,选择 Run As/Run on Server 打开图 2.3 所示的对话框,在对话框中单击 Finish 按钮即完成发布并运行。

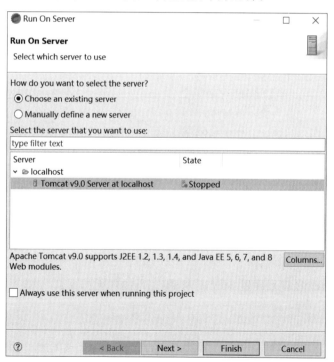

图 2.3 在 STS 中发布并运行 Spring MVC 应用

通过地址 http://localhost:8080/ch2_1 首先访问 index.jsp 页面，如图 2.4 所示。

图 2.4　index.jsp 页面

在图 2.4 所示的页面中，用户单击"注册"超链接时，将请求路径 index/register 与控制器中@RequestMapping 的值对应，找到控制器的请求处理方法 register()，处理后跳转到/WEB-INF/jsp/register.jsp 视图。同理，单击"登录"超链接时，找到控制器的请求处理方法 login()，处理后转到/WEB-INF/jsp/login.jsp 视图。

2.3　基于 Java 配置的 Spring MVC 应用

2.2 节使用 web.xml 和 springmvc-servlet.xml 配置文件进行 Web 配置和 Spring MVC 配置。但本书推荐使用 Java 配置的方式进行项目配置，因此，本节通过一个实例来演示 Spring MVC 应用的 Java 配置。

【例 2-2】Spring MVC 应用的 Java 配置。

该实例的具体要求是：将【例 2-1】中的 springmvc-servlet.xml 和 web.xml 配置文件替换为 Java 配置类。

具体实体步骤如下。

1. 创建 Maven 项目 ch2_2 并添加依赖的 JAR 包

这一步骤与 2.2.1 节相同，不再赘述。

2. 复制 JSP 和 Java 文件

按照相同目录复制应用 ch2_1 中的 JSP 和 Java 文件到应用 ch2_2 中。

3. 创建 Spring MVC 的 Java 配置（相当于 springmvc-servlet.xml 文件）

在 ch2_2 应用的 src/main/java 目录中创建名为 config 的包，在该包中创建 Spring MVC 的 Java 配置类 SpringMVCConfig。在该配置类中使用@Configuration 注解声明该类为 Java 配置类；使用@EnableWebMvc 注解开启默认配置，如 ViewResolver；使用@ComponentScan 注解扫描注解的类；使用@Bean 注解配置视图解析器；该类需要实现 WebMvcConfigurer 接口来配置 Spring MVC。具体代码如下：

```
package config;
import org.springframework.context.annotation.Bean;
import org.springframework.context.annotation.ComponentScan;
import org.springframework.context.annotation.Configuration;
import org.springframework.web.servlet.config.annotation.EnableWebMvc;
```

```java
import org.springframework.web.servlet.config.annotation.WebMvcConfigurer;
import org.springframework.web.servlet.view.InternalResourceViewResolver;
@Configuration
@EnableWebMvc  //开启 Spring MVC 的支持
@ComponentScan("controller")
public class SpringMVCConfig implements WebMvcConfigurer {
    /**
     * 配置视图解析器
     */
    @Bean
    public InternalResourceViewResolver getViewResolver() {
        InternalResourceViewResolver viewResolver =new 
InternalResourceViewResolver();
        viewResolver.setPrefix("/WEB-INF/jsp/");
        viewResolver.setSuffix(".jsp");
        return viewResolver;
    }
}
```

4. 创建 Web 的 Java 配置（相当于 web.xml 文件）

在 ch2_2 应用的 config 包中创建 Web 的 Java 类 WebConfig。该类需要实现 WebApplicationInitializer 接口替代 web.xml 文件的配置。实现该接口将会自动启动 Servlet 容器。在 WebConfig 类中需要使用 AnnotationConfigWebApplicationContext 注册 Spring MVC 的 Java 配置类 SpringMVCConfig，并和当前 ServletContext 关联。最后，在该类中需要注册 Spring MVC 的 DispatcherServlet。具体代码如下：

```java
package config;
import javax.servlet.ServletContext;
import javax.servlet.ServletException;
import javax.servlet.ServletRegistration.Dynamic;
import org.springframework.web.WebApplicationInitializer;
import org.springframework.web.context.support.AnnotationConfigWebApplicationContext;
import org.springframework.web.servlet.DispatcherServlet;
public class WebConfig implements WebApplicationInitializer{
    @Override
    public void onStartup(ServletContext arg0) throws ServletException {
        AnnotationConfigWebApplicationContext ctx
            =new AnnotationConfigWebApplicationContext();
        ctx.register(SpringMVCConfig.class);//注册 Spring MVC 的 Java 配置
                                            //类 SpringMVCConfig
        ctx.setServletContext(arg0);//和当前 ServletContext 关联
        /**
         * 注册 Spring MVC 的 DispatcherServlet
```

```
        */
    Dynamic servlet =arg0.addServlet("dispatcher", new DispatcherServlet(ctx));
    servlet.addMapping("/");
    servlet.setLoadOnStartup(1);
    }
}
```

5. 发布并运行 Spring MVC 应用

选中应用名称 ch2_2,右击,选择 Run As/Run on Server 发布并运行应用。

2.4 实践环节

参考【例 2-1】和【例 2-2】,创建一个 Spring MVC 应用 practice24,分别使用 XML 配置文件和 Java 配置类配置 practice24,并创建 JSP 页面和控制器类测试该 Spring MVC 应用。

2.5 本章小结

本章首先简单介绍了 MVC 的设计模式;其次详细讲解了 Spring MVC 的工作原理;再次,以 ch2_1 应用为例,简要介绍了 Spring MVC 应用的开发步骤;最后以 ch2_2 应用为例,介绍了基于 Java 配置的 Spring MVC 应用的开发步骤。

习 题 2

1. 开发 Spring MVC 应用时,如何配置 DispatcherServlet? 又如何配置 Spring MVC?

2. 简述 Spring MVC 的工作流程。

Spring MVC 的 Controller

学习目的与要求

本章重点讲解 Controller 接收请求参数的方式以及如何编写请求处理方法。通过本章的学习,掌握基于注解的控制器的编写方法,掌握在 Controller 中如何接收请求参数以及编写请求处理方法。

本章主要内容

- 基于注解的控制器。
- 编写请求处理方法。
- Controller 接收请求参数的方式。
- 重定向和转发。
- 应用@Autowired 进行依赖注入。
- @ModelAttribute。

使用 Spring MVC 进行 Web 应用开发时,Controller 是 Web 应用的核心。Controller 实现类包含了对用户请求的处理逻辑,是用户请求和业务逻辑之间的"桥梁",是 Spring MVC 框架的核心部分,负责具体的业务逻辑处理。

3.1 基于注解的控制器

基于注解的控制器,具有如下两个优点:

(1)在基于注解的控制器类中可以编写多个处理方法,进而可以处理多个请求(动作)。这就允许将相关的操作编写在同一个控制器类中,从而减少控制器类的数量,方便以后的维护。

(2)基于注解的控制器不需要在配置文件中或配置类中部署映射,仅需要使用RequestMapping注解类型注解一个方法,进行请求处理。

在Spring MVC中,最重要的两个注解类型是Controller和RequestMapping,本章将重点介绍它们。

3.1.1 @Controller注解类型

在Spring MVC中,使用org.springframework.stereotype.Controller注解类型声明某类的实例是一个控制器。例如,2.2.4小节中的IndexController控制器类。别忘了在Spring MVC的配置文件中使用＜context:component-scan/＞元素(见【例2-1】)或在Spring MVC配置类中使用@ComponentScan注解(见【例2-2】)指定控制器类的基本包,进而扫描所有注解的控制器类。

3.1.2 @RequestMapping注解类型

在基于注解的控制器类中,可以为每个请求编写对应的处理方法。如何将请求与处理方法一一对应呢?需要使用org.springframework.web.bind.annotation.RequestMapping注解类型,将请求与处理方法一一对应。

1. 方法级别注解

方法级别注解示例代码如下:

```
package controller;
import org.springframework.stereotype.Controller;
import org.springframework.web.bind.annotation.RequestMapping;
@Controller
public class IndexController {
    @RequestMapping(value ="/index/login")
    public String login() {
        /**login 代表逻辑视图名称,需要根据Spring MVC配置中
        *internalResourceViewResolver 的前缀和后缀找到对应的物理视图
        */
        return "login";
    }
    @RequestMapping(value ="/index/register")
```

```
        public String register() {
            return "register";
        }
    }
```

上述示例中有两个 RequestMapping 注解语句，它们都作用在处理方法上。注解的 value 属性将请求 URI 映射到方法，value 属性是 RequestMapping 注解的默认属性，如果就一个 value 属性，则可省略该属性。可以使用如下 URL 访问 login 方法（请求处理方法）。

http://localhost:xxx/yyy/index/login

2. 类级别注解

类级别注解示例代码如下：

```
package controller;
import org.springframework.stereotype.Controller;
import org.springframework.web.bind.annotation.RequestMapping;
@Controller
@RequestMapping("/index")
public class IndexController {
    @RequestMapping("/login")
    public String login() {
        return "login";
    }
    @RequestMapping("/register")
    public String register() {
        return "register";
    }
}
```

在类级别注解的情况下，控制器类中的所有方法都将映射为类级别的请求。可以使用如下 URL 访问 login 方法。

http://localhost:xxx/yyy/index/login

为了方便程序维护，建议开发者采用类级别注解，将相关处理放在同一个控制器类中。例如，对商品的增删改查处理方法都可以放在一个名为 GoodsOperate 的控制类中。

3.1.3 编写请求处理方法

在控制器类中，每个请求处理方法可以有多个不同类型的参数，以及一个多种类型的返回结果。

1. 请求处理方法中常出现的参数类型

如果需要在请求处理方法中使用 Servlet API 类型，可以将这些类型作为请求处理方法的参数类型。Servlet API 参数类型示例代码如下：

```
package controller;
import javax.servlet.http.HttpServletRequest;
import javax.servlet.http.HttpSession;
import org.springframework.stereotype.Controller;
import org.springframework.web.bind.annotation.RequestMapping;
@Controller
@RequestMapping("/index")
public class IndexController {
    @RequestMapping("/login")
    public String login(HttpSession session, HttpServletRequest request) {
        session.setAttribute("skey", "session 范围的值");
        request.setAttribute("rkey", "request 范围的值");
        return "login";
    }
}
```

除了 Servlet API 参数类型外，还有输入输出流、表单实体类、注解类型、与 Spring 框架相关的类型等，这些类型在后续章节中再详细介绍。但特别重要的类型是 org.springframework.ui.Model 类型，该类型是一个包含 Map 的 Spring 框架类型。每次调用请求处理方法时，Spring MVC 都将创建 org.springframework.ui.Model 对象。Model 参数类型示例代码如下：

```
package controller;
import org.springframework.stereotype.Controller;
import org.springframework.ui.Model;
import org.springframework.web.bind.annotation.RequestMapping;
@Controller
@RequestMapping("/index")
public class IndexController {
    @RequestMapping("/register")
    public String register(Model model) {
    /* 在视图中可以使用 EL 表达式 ${success} 取出 model 中的值，有关 EL 相关知识，请参考
       本书有关内容。*/
        model.addAttribute("success", "注册成功");
        return "register";
    }
}
```

2. 请求处理方法常见的返回类型

最常见的返回类型就是代表逻辑视图名称的 String 类型，如前面章节中的请求处理方法。除了 String 类型外，还有 Model、View 以及其他任意的 Java 类型。

3.2　Controller 接收请求参数的常见方式

Controller 接收请求参数的方式有很多种,有的适合 get 请求方式,有的适合 post 请求方式,有的两者都适合。下面分别介绍这些方式,读者可根据实际情况选择合适的接收方式。

3.2.1　通过实体 Bean 接收请求参数

通过一个实体 Bean 来接收请求参数,适用于 get 和 post 提交请求方式。需要注意的是,Bean 的属性名称必须与请求参数名称相同。下面通过具体应用 ch3_1 讲解"通过实体 Bean 接收请求参数"。

【例 3-1】　通过实体 Bean 接收请求参数。

应用 ch3_1 的具体要求是:通过应用程序的主页 index.jsp 的超链接进入注册页面 register.jsp 和登录页面 login.jsp,注册成功跳转到登录页面,登录成功跳转到主页面 main.jsp。

具体实现步骤如下。

1. 创建 Maven 项目并添加依赖的 JAR 包

使用 STS 创建一个名为 ch3_1 的 Maven Project,并通过 pom.xml 文件添加项目所依赖的 JAR 包。ch3_1 添加的依赖有 spring-webmvc。pom.xml 文件内容如下:

```xml
<project xmlns="http://maven.apache.org/POM/4.0.0"
    xmlns:xsi="http://www.w3.org/2001/XMLSchema-instance"
    xsi:schemaLocation="http://maven.apache.org/POM/4.0.0
https://maven.apache.org/xsd/maven-4.0.0.xsd">
    <modelVersion>4.0.0</modelVersion>
    <groupId>com.cn.chenheng</groupId>
    <artifactId>ch3_1</artifactId>
    <version>0.0.1-SNAPSHOT</version>
    <packaging>war</packaging>
    <properties>
        <!--spring 版本号 -->
        <spring.version>5.2.3.RELEASE</spring.version>
    </properties>
    <dependencies>
        <dependency>
            <groupId>org.springframework</groupId>
            <artifactId>spring-webmvc</artifactId>
            <version>${spring.version}</version>
        </dependency>
    </dependencies>
```

```
</project>
```

2. 创建视图文件

在应用 ch3_1 的 src/main/webapp/WEB-INF/jsp/目录下创建 register.jsp、login.jsp 和 main.jsp 文件。

register.jsp 的代码如下:

```jsp
<%@page language="java" contentType="text/html; charset=UTF-8" pageEncoding="UTF-8"%>
<%
String path =request.getContextPath();
String basePath = request.getScheme()+"://"+ request.getServerName()+":"+ request.getServerPort()+path+"/";
%>
<!DOCTYPE html>
<html>
<head>
<base href="<%=basePath%>">
<meta charset="UTF-8">
<title>Insert title here</title>
</head>
<body>
<form action="user/register" method="post" name="registForm">
    <table border=1>
        <tr>
            <td>姓名:</td>
            <td>
                <input type="text" name="uname" value="${uname }"/>
            </td>
        </tr>
        <tr>
            <td>密码:</td>
            <td><input type="password"  name="upass"/></td>
        </tr>
        <tr>
            <td>确认密码:</td>
        <td><input type="password"  name="reupass"/></td>
        </tr>
        <tr>
            <td colspan="2" align="center">
                <input type="submit" value="注册" />
            </td>
        </tr>
```

```
        </table>
    </form>
</body>
</html>
```

login.jsp 的代码如下：

```
<%@page language="java" contentType="text/html; charset=UTF-8" pageEncoding="UTF-8"%>
<%
String path=request.getContextPath();
String basePath=request.getScheme()+"://"+request.getServerName()+":"+request.getServerPort()+path+"/";
%>
<!DOCTYPE html>
<html>
<head>
<base href="<%=basePath%>">
<meta charset="UTF-8">
<title>Insert title here</title>
</head>
<body>
    <form action="user/login" method="post">
    <table>
        <tr>
            <td align="center" colspan="2">登录</td>
        </tr>
        <tr>
            <td>姓名:</td>
            <td><input type="text" name="uname"></td>
        </tr>
        <tr>
            <td>密码:</td>
            <td><input type="password" name="upass"></td>
        </tr>
        <tr>
            <td colspan="2">
                <input type="submit" value="提交">
                <input type="reset" value="重置">
            </td>
        </tr>
    </table>
    ${messageError }
    </form>
</body>
```

```
</html>
```

main.jsp 的代码如下：

```jsp
<%@page language="java" contentType="text/html; charset=UTF-8"
    pageEncoding="UTF-8"%>
<%
String path = request.getContextPath();
String basePath = request.getScheme()+"://"+request.getServerName()+":"+
request.getServerPort()+path+"/";
%>
<!DOCTYPE html>
<html>
<head>
<base href="<%=basePath%>">
<meta charset="UTF-8">
<title>Insert title here</title>
</head>
<body>
    欢迎${sessionScope.user.uname}
</body>
</html>
```

在应用 ch3_1 的 src/main/webapp/目录下创建 index.jsp（程序入口页面）页面，代码如下：

```jsp
<%@page language="java" contentType="text/html; charset=UTF-8" pageEncoding="UTF-8"%>
<%
String path = request.getContextPath();
String basePath = request.getScheme()+"://"+request.getServerName()+":"+
request.getServerPort()+path+"/";
%>
<!DOCTYPE html>
<html>
<head>
<base href="<%=basePath%>">
<meta charset="UTF-8">
<title>Insert title here</title>
</head>
<body>
    没注册的用户,请<a href="user/toRegister">注册</a>!<br>
    已注册的用户,去<a href="user/toLogin">登录</a>!
</body>
</html>
```

3. 创建 POJO 实体类

在应用 ch3_1 的 src/main/java 目录下创建一个名为 pojo 的包,并在该包中创建实体类 UserForm,代码如下:

```java
package pojo;
public class UserForm {
    private String uname;//与请求参数名称相同
    private String upass;
    private String reupass;
    //省略 getter 和 setter 方法
}
```

4. 创建控制器类

在应用 ch3_1 的 src/main/java 目录下创建一个名为 controller 的包,并在该包中创建控制器类 UserController。

UserController 的代码如下:

```java
package controller;
import javax.servlet.http.HttpSession;
import org.apache.commons.logging.Log;
import org.apache.commons.logging.LogFactory;
import org.springframework.stereotype.Controller;
import org.springframework.ui.Model;
import org.springframework.web.bind.annotation.RequestMapping;
import pojo.UserForm;
@Controller
@RequestMapping("/user")
public class UserController {
    //得到一个用来记录日志的对象,这样打印信息时能够标记是打印的哪个类的信息
    private static final Log logger =LogFactory.getLog(UserController.class);
    /**
     * 从首页跳转到注册页面
     */
    @RequestMapping("/toRegister")
    public String toRegister() {
        return "register";
    }
    /**
     * 从首页跳转到登录页面
     */
    @RequestMapping("/toLogin")
    public String toLogin() {
```

```java
            return "login";
    }
    /**
     * 处理登录
     * 使用UserForm对象(实体bean)user接收登录页面提交的请求参数
     */
    @RequestMapping("/login")
    public String login(UserForm user, HttpSession session, Model model) {
        if("zhangsan".equals(user.getUname())
                && "123456".equals(user.getUpass())) {
            session.setAttribute("user", user);
            logger.info("成功");
            return "main";//登录成功,跳转到main.jsp
        }else{
            logger.info("失败");
            model.addAttribute("messageError", "用户名或密码错误");
            return "login";
        }
    }
    /**
     * 处理注册
     * 使用UserForm对象(实体bean)user接收注册页面提交的请求参数
     */
    @RequestMapping("/register")
    public String register(UserForm user, Model model) {
        if("zhangsan".equals(user.getUname())
                && "123456".equals(user.getUpass())) {
            logger.info("成功");
            return "login";//注册成功,跳转到login.jsp
        }else{
            logger.info("失败");
            //在register.jsp页面上可以使用EL表达式取出model的uname值
            model.addAttribute("uname", user.getUname());
            return "register";//返回register.jsp
        }
    }
}
```

5. 创建Spring MVC的Java配置(相当于springmvc-servlet.xml文件)

在ch3_1应用的src/main/java目录中创建名为config的包,在该包中创建Spring MVC的Java配置类SpringMVCConfig。在该配置类中使用@Configuration注解声明该类为Java配置类;使用@EnableWebMvc注解开启默认配置,如ViewResolver;使用@ComponentScan注解扫描注解的类;使用@Bean注解配置视图解析器和静态资源;该类需要实现 WebMvcConfigurer 接口来配置 Spring MVC。具体代码如下：

```
package config;
import org.springframework.context.annotation.Bean;
import org.springframework.context.annotation.ComponentScan;
import org.springframework.context.annotation.Configuration;
import org.springframework.web.servlet.config.annotation.EnableWebMvc;
import org.springframework.web.servlet.config.annotation.ResourceHandlerRegistry;
import org.springframework.web.servlet.config.annotation.WebMvcConfigurer;
import org.springframework.web.servlet.view.InternalResourceViewResolver;
@Configuration
@EnableWebMvc
@ComponentScan(basePackages = {"controller","service"})//扫描基本包
public class SpringMVCConfig implements WebMvcConfigurer {
    /**
     * 配置视图解析器
     */
    @Bean
    public InternalResourceViewResolver getViewResolver() {
        InternalResourceViewResolver viewResolver = new InternalResourceViewResolver();
        viewResolver.setPrefix("/WEB-INF/jsp/");
        viewResolver.setSuffix(".jsp");
        return viewResolver;
    }
    /**
     * 配置静态资源(不需要 DispatcherServlet 转发的请求)
     */
    @Override
    public void addResourceHandlers(ResourceHandlerRegistry registry) {
        registry.addResourceHandler("/static/**").addResourceLocations("/static/");
    }
}
```

6. 创建 Web 的 Java 配置（相当于 web.xml 文件）

在 ch3_1 应用的 config 包中创建 Web 的 Java 类 WebConfig。该类需要实现 WebApplicationInitializer 接口替代 web.xml 文件的配置。实现该接口将会自动启动 Servlet 容器。在 WebConfig 类中需要使用 AnnotationConfigWebApplicationContext 注册 Spring MVC 的 Java 配置类 SpringMVCConfig，并和当前 ServletContext 关联。最后，在该类中需要注册 Spring MVC 的 DispatcherServlet。具体代码如下：

```
package config;
import javax.servlet.ServletContext;
import javax.servlet.ServletException;
import javax.servlet.ServletRegistration.Dynamic;
import org.springframework.web.WebApplicationInitializer;
import org.springframework.web.context.support.AnnotationConfigWebApplica-
tionContext;
import org.springframework.web.servlet.DispatcherServlet;
public class WebConfig implements WebApplicationInitializer{
    @Override
    public void onStartup(ServletContext arg0) throws ServletException {
        AnnotationConfigWebApplicationContext ctx
        =new AnnotationConfigWebApplicationContext();
        ctx.register(SpringMVCConfig.class);//注册 Spring MVC 的 Java 配置
                                            //类 SpringMVCConfig
        ctx.setServletContext(arg0);//和当前 ServletContext 关联
        /**
         * 注册 Spring MVC 的 DispatcherServlet
         */
        Dynamic servlet = arg0.addServlet("dispatcher", new DispatcherServlet
                                            (ctx));
        servlet.addMapping("/");
        servlet.setLoadOnStartup(1);
    }
}
```

7. 发布并测试应用

选中应用名称 ch3_1,右击,选择 Run As/Run on Server 发布并测试应用。

3.2.2 通过处理方法的形参接收请求参数

通过处理方法的形参接收请求参数,也就是直接把表单参数写在控制器类相应方法的形参中,即形参名称与请求参数名称完全相同。该接收参数方式适用于 get 和 post 提交请求方式。可以将 3.2.1 节中控制器类 UserController 中 register 方法的代码修改如下:

```
@RequestMapping("/register")
/**
 *通过形参接收请求参数,形参名称与请求参数名称完全相同
 */
public String register(String uname, String upass, Model model) {
    if("zhangsan".equals(uname)
        && "123456".equals(upass)) {
        logger.info("成功");
```

```
        return "login";//注册成功,跳转到login.jsp
    }else{
        logger.info("失败");
        //在register.jsp页面上可以使用EL表达式取出model的uname值
        model.addAttribute("uname", uname);
        return "register";//返回register.jsp
    }
}
```

3.2.3 通过 HttpServletRequest 接收请求参数

通过 HttpServletRequest 接收请求参数,适用于 get 和 post 提交请求方式。可以将 3.2.1 节中控制器类 UserController 中 register 方法的代码修改如下:

```
@RequestMapping("/register")
/*
 * 通过HttpServletRequest 接收请求参数
 */
public String register(HttpServletRequest request, Model model) {
    String uname = request.getParameter("uname");
    String upass = request.getParameter("upass");
    if("zhangsan".equals(uname)
            && "123456".equals(upass)) {
        logger.info("成功");
        return "login";//注册成功,跳转到login.jsp
    }else{
        logger.info("失败");
        //在register.jsp页面上可以使用EL表达式取出model的uname值
        model.addAttribute("uname", uname);
        return "register";//返回register.jsp
    }
}
```

3.2.4 通过@PathVariable 接收 URL 中的请求参数

通过@PathVariable 获取 URL 中的参数,控制器类示例代码如下:

```
package controller;
import org.springframework.stereotype.Controller;
import org.springframework.ui.Model;
import org.springframework.web.bind.annotation.PathVariable;
import org.springframework.web.bind.annotation.RequestMapping;
import org.springframework.web.bind.annotation.RequestMethod;
@Controller
@RequestMapping("/user")
```

```java
public class UserController {
    @RequestMapping(value="/register/{uname}/{upass}", method=RequestMethod.GET)
    //必须加 method 属性
    /**
     * 通过@PathVariable 获取 URL 中的参数
     */
    public String register(@PathVariable String uname, @PathVariable String
            upass, Model model) {
        if("zhangsan".equals(uname)
                && "123456".equals(upass))
            return "login";//注册成功,跳转到 login.jsp
        else{
            //在 register.jsp 页面可以使用 EL 表达式取出 model 的 uname 值
            model.addAttribute("uname", uname);
            return "register";//返回 register.jsp
        }
    }
}
```

访问 http://localhost:8080/ch3_1/user/register/zhangsan/123456 路径时,上述代码自动将 URL 中模板变量{uname}和{upass}绑定到通过@PathVariable 注解的同名参数上,即 uname=zhangsan,upass=123456。

3.2.5 通过@RequestParam 接收请求参数

通过@RequestParam 接收请求参数,适用于 get 和 post 提交请求方式。可以将 3.2.1 节中控制器类 UserController 中 register 方法的代码修改如下:

```java
@RequestMapping("/register")
/**
 * 通过@RequestParam 接收请求参数
 */
public String register(@RequestParam String uname, @RequestParam String upass,
        Model model) {
    if("zhangsan".equals(uname)
            && "123456".equals(upass)) {
        logger.info("成功");
        return "login";//注册成功,跳转到 login.jsp
    }else{
        logger.info("失败");
        //在 register.jsp 页面可以使用 EL 表达式取出 model 的 uname 值
        model.addAttribute("uname", uname);
        return "register";//返回 register.jsp
    }
}
```

通过@RequestParam 接收请求参数与 3.2.2 节"通过处理方法的形参接收请求参数"的区别是：当请求参数与接收参数名不一致时，"通过处理方法的形参接收请求参数"不会报 400 错误，而"通过@RequestParam 接收请求参数"会报 400 错误。

3.2.6 通过@ModelAttribute 接收请求参数

@ModelAttribute 注解放在处理方法的形参上时，用于将多个请求参数封装到一个实体对象，从而简化数据绑定流程，而且自动暴露为模型数据，在视图页面展示时使用。而 3.2.1 节只是将多个请求参数封装到一个实体对象，并不能暴露为模型数据（需要使用 model.addAttribute 语句才能暴露为模型数据，数据绑定与模型数据展示，可参考第 5 章的内容）。

通过@ModelAttribute 注解接收请求参数，适用于 get 和 post 提交请求方式。可以将 3.2.1 节中控制器类 UserController 中 register 方法的代码修改如下：

```
@RequestMapping("/register")
public String register(@ModelAttribute("user") UserForm user) {
    if("zhangsan".equals(user.getUname())
            && "123456".equals(user.getUpass())){
        logger.info("成功");
        return "login";//注册成功，跳转到 login.jsp
    }else{
        logger.info("失败");
    //使用@ModelAttribute("user")与 model.addAttribute("user", user)功能相同
    //在 register.jsp 页面上可以使用 EL 表达式${user.uname}取出 ModelAttribute 的
    //uname 值
        return "register";//返回 register.jsp
    }
}
```

3.3 重定向与转发

重定向是将用户从当前处理请求定向到另一个视图（如 JSP）或处理请求，以前的请求（request）中存放的信息全部失效，并进入一个新的 request 作用域；转发是将用户对当前处理的请求转发给另一个视图或处理请求，以前 request 中存放的信息不会失效。

转发是服务器行为，重定向是客户端行为。具体工作流程如下：

转发过程：客户端浏览器发送 HTTP 请求，Web 服务器接受此请求，调用内部的一个方法，在容器内部完成请求处理和转发动作，将目标资源发送给客户端；在这里，转发的路径必须是同一个 Web 容器下的 URL，其不能转向到其他的 Web 路径上去，中间传递的是自己容器内的 request。客户端浏览器的地址栏中显示的仍然是其第一次访问的路径，也就是说客户端是感觉不到服务器做了转发的。转发行为是浏览器只做了一次访问

请求。

重定向过程：客户端浏览器发送 HTTP 请求，Web 服务器接受后发送 302 状态码响应及对应新的 location 给客户端浏览器，客户端浏览器发现是 302 响应，则自动再发送一个新的 HTTP 请求，请求 URL 是新的 location 地址，服务器根据此请求寻找资源，并发送给客户端。这里 location 可以重定向到任意 URL，既然是浏览器重新发出了请求，则就没有什么 request 传递的概念了。客户端浏览器的地址栏中显示的是其重定向的路径，客户端可以观察到地址的变化。重定向行为是浏览器做了至少两次的访问请求。

在 Spring MVC 框架中，控制器类中处理方法的 return 语句默认就是转发实现，只不过实现的是转发到视图。示例代码如下：

```
@RequestMapping("/register")
public String register() {
    return "register";//转发到 register.jsp
}
```

在 Spring MVC 框架中，重定向与转发的示例代码如下：

```
package controller;
import org.springframework.stereotype.Controller;
import org.springframework.web.bind.annotation.RequestMapping;
@Controller
@RequestMapping("/index")
public class IndexController {
    @RequestMapping("/login")
    public String login() {
        //转发到一个请求方法(同一个控制器类里,可省略/index/)
        return "forward:/index/isLogin";
    }
    @RequestMapping("/isLogin")
    public String isLogin() {
        //重定向到一个请求方法
        return "redirect:/index/isRegister";
    }
    @RequestMapping("/isRegister")
    public String isRegister() {
        //转发到一个视图
        return "register";
    }
}
```

在 Spring MVC 框架中，不管重定向还是转发，都需要符合视图解析器的配置，如果直接重定向到一个不需要 DispatcherServlet 的资源，如：

```
return "redirect:/html/my.html";
```

在 Spring MVC 配置文件中,使用 mvc:resources 元素配置,示例代码如下:

```
<mvc:annotation-driven />
    <!--annotation-driven 用于简化开发的配置,
    注解 DefaultAnnotationHandlerMapping 和 AnnotationMethodHandlerAdapter -->
    <!--使用 resources 过滤掉不需要 dispatcher servlet 的资源(即静态资源,如 CSS、
JS、HTML、images)。使用 resources 时,必须使用 annotation-driven,不然 resources 元
素会阻止任意控制器被调用。-->
<mvc:resources location="/html/" mapping="/html/ * * "></mvc:resources>
```

在 Spring MVC 配置类中,需要实现 WebMvcConfigurer 的接口方法 public void addResourceHandlers(ResourceHandlerRegistry registry),示例代码如下:

```
@Override
public void addResourceHandlers(ResourceHandlerRegistry registry) {
    registry.addResourceHandler("/html/**").addResourceLocations("/html/");
}
```

3.4 应用@Autowired 进行依赖注入

在前面的控制器中,并没有体现 MVC 的 M 层,这是因为控制器既充当 C 层,又充当 M 层。这样设计程序的系统结构很不合理,应该将 M 层从控制器中分离出来。Spring MVC 框架本身就是一个非常优秀的 MVC 框架,它具有一个依赖注入的优点。可以通过 org.springframework.
beans.factory.annotation.Autowired 注解类型将依赖注入一个属性(成员变量)或方法,如:

```
@Autowired
public UserService userService;
```

在 Spring MVC 中,为了能被作为依赖注入,实现类必须使用 org.springframework. stereotype.Service 注解类型注明为@Service(一个服务)。另外,还需要在配置文件中使用<context:component-scan base-package="基本包"/>元素,或者在配置类中使用 @ComponentScan("基本包")注解来扫描依赖基本包。下面将【例 3-1】的 ch3_1 应用的 "登录"和"注册"的业务逻辑处理分离出来,使用 Service 层实现。

首先,创建 service 包,在包中创建 UserService 接口和 UserServiceImpl 实现类。
UserService 接口的具体代码如下:

```
package service;
import pojo.UserForm;
public interface UserService {
    boolean login(UserForm user);
    boolean register(UserForm user);
}
```

UserServiceImpl 实现类的具体代码如下：

```java
package service;
import org.springframework.stereotype.Service;
import pojo.UserForm;
//注解为一个服务
@Service
public class UserServiceImpl implements UserService{
    @Override
    public boolean login(UserForm user) {
        if("zhangsan".equals(user.getUname())
                && "123456".equals(user.getUpass()))
            return true;
        return false;
    }
    @Override
    public boolean register(UserForm user) {
        if("zhangsan".equals(user.getUname())
                && "123456".equals(user.getUpass()))
            return true;
        return false;
    }
}
```

其次，将配置类 SpringMVCConfig 的@ComponentScan("controller")语句修改如下：

```java
@ComponentScan(basePackages ={"controller","service"})//扫描基本包
```

最后，修改控制器类 UserController，具体代码如下：

```java
package controller;
import javax.servlet.http.HttpSession;
import org.apache.commons.logging.Log;
import org.apache.commons.logging.LogFactory;
import org.springframework.beans.factory.annotation.Autowired;
import org.springframework.stereotype.Controller;
import org.springframework.ui.Model;
import org.springframework.web.bind.annotation.RequestMapping;
import pojo.UserForm;
import service.UserService;
@Controller
@RequestMapping("/user")
public class UserController {
    //得到一个用来记录日志的对象，这样打印信息时能够标记打印的是哪个类的信息
    private static final Log logger =LogFactory.getLog(UserController.class);
    //将服务依赖注入属性 userService
    @Autowired
```

```java
    public UserService userService;
    /**
     * 处理登录
     */
    @RequestMapping("/login")
    public String login(UserForm user, HttpSession session, Model model) {
        if(userService.login(user)){
            session.setAttribute("u", user);
            logger.info("成功");
            return "main";//登录成功,跳转到main.jsp
        }else{
            logger.info("失败");
            model.addAttribute("messageError", "用户名或密码错误");
            return "login";
        }
    }
    /**
     *处理注册
     */
    @RequestMapping("/register")
    public String register(@ModelAttribute("user") UserForm user) {
        if(userService.register(user)){
            logger.info("成功");
            return "login";//注册成功,跳转到login.jsp
        }else{
            logger.info("失败");
            //使用@ModelAttribute("user")与model.addAttribute("user", user)功
              能相同
        //在register.jsp页面上可以使用EL表达式${user.uname}取出ModelAttribute的
          uname值
            return "register";//返回register.jsp
        }
    }
}
```

3.5　@ModelAttribute

通过org.springframework.web.bind.annotation.ModelAttribute注解类型,可经常实现如下两个功能。

1. 绑定请求参数到实体对象(表单的命令对象)

该用法如 3.2.6 节内容代码如下:

```java
@RequestMapping("/register")
public String register(@ModelAttribute("user") UserForm user) {
    if("zhangsan".equals(user.getUname())
            && "123456".equals(user.getUpass())){
        return "login";
    }else{
        return "register";
    }
}
```

上述代码中@ModelAttribute("user") UserForm user 语句的功能有两个：一是将请求参数的输入封装到 user 对象中；二是创建 UserForm 实例，以 user 为键值存储在 Model 对象中，与 model.addAttribute("user"，user)语句功能一样。如果没有指定键值，即@ModelAttribute UserForm user，那么创建 UserForm 实例时，以 userForm 为键值存储在 Model 对象中，与 model.addAttribute("userForm"，user)语句功能一样。

2. 注解一个非请求处理方法

在控制器类中，被@ModelAttribute 注解的一个非请求处理方法将在每次调用该控制器类的请求处理方法前被调用。这种特性可以用来控制登录权限，当然，控制登录权限的方法很多，例如拦截器、过滤器等。

使用该特性控制登录权限的示例代码如下：

```java
package controller;
import javax.servlet.http.HttpSession;
import org.springframework.web.bind.annotation.ModelAttribute;
public class BaseController {
    @ModelAttribute
    public void isLogin(HttpSession session) throws Exception {
        if(session.getAttribute("user") ==null){
            throw new Exception("没有权限");
        }
    }
}
package controller;
import org.springframework.stereotype.Controller;
import org.springframework.web.bind.annotation.RequestMapping;
@Controller
@RequestMapping("/admin")
public class ModelAttributeController extends BaseController{
    @RequestMapping("/add")
    public String add(){
        return "addSuccess";
    }
```

```
    @RequestMapping("/update")
    public String update(){
        return "updateSuccess";
    }
    @RequestMapping("/delete")
    public String delete(){
        return "deleteSuccess";
    }
}
```

上述 ModelAttributeController 类中的 add、update、delete 请求处理方法执行时,首先执行父类 BaseController 中的 isLogin 方法判断登录权限。

3.6 实 践 环 节

编写一个 Spring MVC 应用 practice36。该应用的具体要求是:通过程序入口 login.jsp 输入用户名和密码,单击"提交"按钮,将输入的信息提交给 LoginController 控制器类。在 LoginController 控制器类中进行登录验证:如果输入正确(用户名 zhangsan,密码 123),转发到 main.jsp 页面,并在 main.jsp 页面显示用户名;如果输入不正确,重定向到 login.jsp 页面。

3.7 本 章 小 结

本章是整个 Spring MVC 框架的核心部分。通过本章的学习,务必掌握如何编写基于注解的控制器类。

习 题 3

1. 在 Spring MVC 的控制器类中如何访问 Servlet API?
2. 控制器接收请求参数的常见方式有哪几种?
3. 如何编写基于注解的控制器类?
4. @ModelAttribute 可实现哪些功能?

类型转换和格式化

学习目的与要求

本章主要学习类型转换器和格式化转换器。通过本章的学习,应该理解类型转换器和格式化转换器的原理,掌握类型转换器和格式化转换器的用法。

本章主要内容

- Converter。
- Formatter。

在 Spring MVC 框架中,需要收集用户请求参数,并将请求参数传递给应用的控制器组件。此时存在一个问题,所有的请求参数类型只能是字符串数据类型,但 Java 是强类型语言,所以 Spring MVC 框架必须将这些字符串请求参数转换成相应的数据类型。

Spring MVC 框架不仅提供了强大的类型转换和格式化机制,而且开发者还可以方便地开发出自己的类型转换器和格式化转换器,完成字符串和各种数据类型之间的转换。这正是学习本章的目的所在。

4.1 类型转换的意义

本节通过一个简单应用(JSP + Servlet)为示例介绍类型转换的意义。如图 4.1 所示的添加商品信息页面用于收集用户输入的商品信息。商品信息包括：商品名称(字符串类型 String)、商品价格(双精度浮点类型 double)、商品数量(整数类型 int)。

图 4.1 添加商品信息的收集页面

addGoods.jsp 页面的代码如下：

```
<body>
    <form action="addGoods" method="post">
        商品名称：<input type="text" name="goodsname"/><br>
        商品价格：<input type="text" name="goodsprice"/><br>
        商品数量：<input type="text" name="goodsnumber"/><br>
        <input type="submit" value="提交"/>
    </form>
</body>
```

希望页面收集到的数据提交到 addGoods 的 Servlet(AddGoodsServlet 类)，该 Servlet 将这些请求信息封装成一个 Goods 类的值对象。

Goods 类的代码如下：

```
package pojo;
public class Goods {
    private String goodsname;
    private double goodsprice;
    private int goodsnumber;
    //无参数的构造方法
    public Goods(){}
    //有参数的构造方法
    public Goods(String goodsname, double goodsprice, int goodsnumber) {
        super();
        this.goodsname = goodsname;
        this.goodsprice = goodsprice;
        this.goodsnumber = goodsnumber;
    }
    //此处省略了 setter 和 getter 方法
}
```

AddGoodsServlet 类的代码如下：

```
package servlet;
import java.io.IOException;
import javax.servlet.ServletException;
import javax.servlet.http.HttpServlet;
import javax.servlet.http.HttpServletRequest;
import javax.servlet.http.HttpServletResponse;
import domain.Goods;
public class AddGoodsServlet extends HttpServlet {
    public void doGet(HttpServletRequest request, HttpServletResponse response)
throws ServletException, IOException {
        doPost(request, response);
    }
    public void doPost(HttpServletRequest request, HttpServletResponse response)
            throws ServletException, IOException {
        response.setContentType("text/html;charset=utf-8");
        //设置编码,防止乱码
        request.setCharacterEncoding("utf-8");
        //获取参数值
        String goodsname = request.getParameter("goodsname");
        String goodsprice = request.getParameter("goodsprice");
        String goodsnumber = request.getParameter("goodsnumber");
        //下面进行类型转换
        double newgoodsprice = Double.parseDouble(goodsprice);
        int newgoodsnumber = Integer.parseInt(goodsnumber);
        //将转换后的数据封装成 goods 值对象
        Goods goods = new Goods(goodsname, newgoodsprice, newgoodsnumber);
        //将 goods 值对象传递给数据访问层,进行添加操作,代码省略
    }
}
```

对于上面这个应用而言,开发者需要自己在 Servlet 中进行类型转换,并将其封装成值对象。这些类型转换操作全部手工完成,异常烦琐。

对于 Spring MVC 框架而言,它必须将请求参数转换成值对象类里各属性对应的数据类型——这就是类型转换的意义。

4.2 Converter

Spring MVC 框架的 Converter<S,T>是一个可以将一种数据类型转换成另一种数据类型的接口,这里 S 表示源类型,T 表示目标类型。在实际应用中,开发者使用框架内置的类型转换器基本就够了,但有时需要编写具有特定功能的类型转换器。

4.2.1 内置的类型转换器

在 Spring MVC 框架中，对于常用的数据类型，开发者无须创建自己的类型转换器，因为 Spring MVC 框架有许多内置的类型转换器，以完成常用的类型转换。Spring MVC 框架提供的内置类型转换器包括如下几种类型。

（1）标量转换器。

StringToBooleanConverter：String 到 Boolean 类型转换。

ObjectToStringConverter：Object 到 String 转换，调用 toString 方法转换。

StringToNumberConverterFactory：String 到数字转换（如 Integer、Long 等）。

NumberToNumberConverterFactory：数字子类型（基本类型）到数字类型（包装类型）转换。

StringToCharacterConverter：String 到 Character 转换，取字符串第一个字符。

NumberToCharacterConverter：数字子类型到 Character 转换。

CharacterToNumberFactory：Character 到数字子类型转换。

StringToEnumConverterFactory：String 到枚举类型转换，通过 Enum.valueOf 将字符串转换为需要的枚举类型。

EnumToStringConverter：枚举类型到 String 转换，返回枚举对象的 name() 值。

StringToLocaleConverter：String 到 java.util.Locale 转换。

PropertiesToStringConverter：java.util.Properties 到 String 转换，默认通过 ISO-8859-1 解码。

StringToPropertiesConverter：String 到 java.util.Properties 转换，默认使用 ISO-8859-1 编码。

（2）集合、数组相关转换器。

ArrayToCollectionConverter：任意数组到任意集合（List、Set）转换。

CollectionToArrayConverter：任意集合到任意数组转换。

ArrayToArrayConverter：任意数组到任意数组转换。

CollectionToCollectionConverter：集合之间的类型转换。

MapToMapConverter：Map 之间的类型转换。

ArrayToStringConverter：任意数组到 String 转换。

StringToArrayConverter：字符串到数组的转换，默认通过","分隔，且去除字符串的两边空格（trim）。

ArrayToObjectConverter：任意数组到 Object 的转换，如果目标类型和源类型兼容，直接返回源对象；否则返回数组的第一个元素，并进行类型转换。

ObjectToArrayConverter：Object 到单元素数组转换。

CollectionToStringConverter：任意集合（List、Set）到 String 转换。

StringToCollectionConverter：String 到集合（List、Set）转换，默认通过","分隔，且去除字符串的两边空格（trim）。

CollectionToObjectConverter：任意集合到任意 Object 的转换，如果目标类型和源

类型兼容,直接返回源对象;否则返回集合的第一个元素,并进行类型转换。

ObjectToCollectionConverter:Object 到单元素集合的类型转换。

类型转换是在视图与控制器相互传递数据时发生的。Spring MVC 框架对于基本类型,如 int、long、float、double、boolean 以及 char 等,已经做好了基本类型转换。例如,针对 4.1 节 addGoods.jsp 的提交请求,可以由如下处理方法来接收请求参数并处理:

```
package controller;
import org.springframework.stereotype.Controller;
import org.springframework.web.bind.annotation.RequestMapping;
@Controller
public class GoodsController {
    @RequestMapping("/addGoods")
    public String add(String goodsname, double goodsprice, int goodsnumber){
        double total =goodsprice * goodsnumber;
        System.out.println(total);
        return "success";
    }
}
```

注意:使用内置类型转换器时,请求参数输入值与接收参数类型要兼容,否则报 400 错误。请求参数类型与接收参数类型不兼容问题,需要学习输入校验后才可解决。

4.2.2 自定义类型转换器

当 Spring MVC 框架内置的类型转换器不能满足需求时,开发者可以开发自己的类型转换器。例如,有个应用 ch4_1 希望用户在页面表单中输入信息来创建商品信息。当输入 apple,10.58,200 时,表示在程序中自动创建一个 new Goods,并将 apple 值自动赋值给 goodsname 属性,将 10.58 的值自动赋值给 goodsprice 属性,将 200 的值自动赋值给 goodsnumber 属性。

欲实现上述应用,需要完成以下 5 项工作:

① 创建相关视图。
② 创建实体类。
③ 创建控制器类。
④ 创建自定义类型转换器。
⑤ 注册类型转换器。

【例 4-1】 按照上述步骤,采用自定义类型转换器完成应用 ch4_1 的需求。

具体实现步骤如下。

1. 创建 Maven 项目并添加依赖的 JAR 包

使用 STS 创建一个名为 ch4_1 的 Maven Project,并通过 pom.xml 文件添加项目所依赖的 JAR 包。ch4_1 添加的依赖有 spring-webmvc。pom.xml 文件内容与【例 3-1】的相同,不再赘述。

2. 创建相关视图

在 ch4_1 的 src/main/webapp 目录下创建信息采集页面 input.jsp,代码如下:

```jsp
<%@page language="java" contentType="text/html; charset=UTF-8"
    pageEncoding="UTF-8"%>
<%
String path =request.getContextPath();
String basePath = request.getScheme()+"://"+request.getServerName()+":"+
request.getServerPort()+path+"/";
%>
<!DOCTYPE html>
<html>
<head>
<base href="<%=basePath%>">
<meta charset="UTF-8">
<title>Insert title here</title>
</head>
<body>
    <form action="my/converter" method="post">
        请输入商品信息(格式为:apple,10.58,200):<input type="text" name="goods" />
<br>
        <input type="submit" value="提交" />
    </form>
</body>
</html>
```

在 ch4_1 的 src/main/webapp/WEB-INF/jsp/ 目录下创建信息显示页面 showGoods.jsp, 代码如下:

```jsp
<%@page language="java" contentType="text/html; charset=UTF-8"
    pageEncoding="UTF-8"%>
<%
String path =request.getContextPath();
String basePath = request.getScheme()+"://"+request.getServerName()+":"+
request.getServerPort()+path+"/";
%>
<!DOCTYPE html>
<html>
<head>
<base href="<%=basePath%>">
<meta charset="UTF-8">
<title>Insert title here</title>
</head>
<body>
```

```
    您创建的商品信息如下:<br>
    <!--使用 EL 表达式取出 model 中 goods 的信息 -->
    商品名为:${goods.goodsname }?
    商品价格为:${goods.goodsprice }?
    商品数量为:${goods.goodsnumber }?
</body>
</html>
```

3. 创建实体类

在 ch4_1 的 src/main/java 目录下创建一个名为 pojo 的包,并在该包中创建名为 GoodsModel 的实体类,代码如下:

```
package pojo;
public class GoodsModel {
    private String goodsname;
    private double goodsprice;
    private int goodsnumber;
    //省略 setter 和 getter 方法
}
```

4. 创建控制器类

在 ch4_1 的 src/main/java 目录下创建一个名为 controller 的包,并在该包中创建名为 ConverterController 的控制器类,代码如下:

```
package controller;
import org.springframework.stereotype.Controller;
import org.springframework.ui.Model;
import org.springframework.web.bind.annotation.RequestMapping;
import org.springframework.web.bind.annotation.RequestParam;
import pojo.GoodsModel;
@Controller
@RequestMapping("/my")
public class ConverterController {
    @RequestMapping("/converter")
    /*使用@RequestParam("goods")接收请求参数,
    然后调用自定义类型转换器 GoodsConverter 将字符串值转换为 GoodsModel 的对象 gm
    */
    public String myConverter(@RequestParam("goods") GoodsModel gm, Model model){
        model.addAttribute("goods",gm);
        return "showGoods";
    }
}
```

5. 创建自定义类型转换器

自定义类型转换器类需要实现 Converter<S，T>接口，重写 convert(S)接口方法。convert(S)方法功能是将源数据类型 S 转换成目标数据类型 T。

在 ch4_1 的 src/main/java 目录下，创建一个名为 converter 的包，并在该包中创建名为 GoodsConverter 的自定义类型转换器类。该类是一个组件类，需要使用@Component 注解标注。具体代码如下：

```java
package converter;
import org.springframework.core.convert.converter.Converter;
import org.springframework.stereotype.Component;
import pojo.GoodsModel;
@Component
public class GoodsConverter implements Converter<String, GoodsModel>{
    @Override
    public GoodsModel convert(String source) {
        // 创建一个 Goods 实例
        GoodsModel goods =new GoodsModel();
        // 以","分隔的
        String stringValues[] =source.split(",");
        if (stringValues !=null && stringValues.length ==3) {
            // 为 Goods 实例赋值
            goods.setGoodsname(stringValues[0]);
            goods.setGoodsprice(Double.parseDouble(stringValues[1]));
            goods.setGoodsnumber(Integer.parseInt(stringValues[2]));
            return goods;
        } else {
            throw new IllegalArgumentException(String.format("类型转换失败,需要格式'apple,10.58,200',但格式是[%s]", source));
        }
    }
}
```

6. 创建配置类，并注册类型转换器

在 ch4_1 应用的 src/main/java 目录中创建一个名为 config 的包，并在该包中创建 Spring MVC 的 Java 配置类 SpringMVCConfig。在该配置类中使用@Configuration 注解声明该类为 Java 配置类；使用@EnableWebMvc 注解开启默认配置，如 ViewResolver；使用@ComponentScan 注解扫描注解的类；使用@Bean 注解配置视图解析器；该类需要实现 WebMvcConfigurer 接口来配置 Spring MVC；使用 ConfigurableConversionService 接口注册类型转换器。具体代码如下：

```java
package config;
```

```java
import javax.annotation.PostConstruct;
import org.springframework.beans.factory.annotation.Autowired;
import org.springframework.context.annotation.Bean;
import org.springframework.context.annotation.ComponentScan;
import org.springframework.context.annotation.Configuration;
import org.springframework.core.convert.support.ConfigurableConversionService;
import org.springframework.web.servlet.config.annotation.EnableWebMvc;
import org.springframework.web.servlet.config.annotation.WebMvcConfigurer;
import org.springframework.web.servlet.view.InternalResourceViewResolver;
import converter.GoodsConverter;
@Configuration
@EnableWebMvc // 开启 Spring MVC 的支持
@ComponentScan(basePackages = { "controller", "converter" }) // 扫描基本包
public class SpringMVCConfig implements WebMvcConfigurer {
    @Autowired
    private ConfigurableConversionService conversionService;// 用于注册类型转换器
    @Autowired
    private GoodsConverter goodsConverter;
    /**
     * 配置视图解析器
     */
    @Bean
    public InternalResourceViewResolver getViewResolver() {
        InternalResourceViewResolver viewResolver =new
InternalResourceViewResolver();
        viewResolver.setPrefix("/WEB-INF/jsp/");
        viewResolver.setSuffix(".jsp");
        return viewResolver;
    }
    /**
     * 注册类型转换器
     */
    @PostConstruct
    /**
     * 被@PostConstruct 修饰的方法会在服务器加载 Servlet 的时候运行,并且只会被服务
       器调用一次,类似于 Servlet 的 inti()方法。
     * 被@PostConstruct 修饰的方法会在构造函数之后,init()方法之前运行。
     */
    public void initEditableAvlidation() {
        if (conversionService !=null) {
            conversionService.addConverter(goodsConverter);
            //这里可以注册多个类型转换器
        }
    }
}
```

在 ch4_1 应用的 config 包中创建 Web 的 Java 类 WebConfig。该类需要实现 WebApplicationInitializer 接口替代 web.xml 文件的配置。实现该接口将会自动启动 Servlet 容器。在 WebConfig 类中需要使用 AnnotationConfigWebApplicationContext 注册 Spring MVC 的 Java 配置类 SpringMVCConfig，并和当前 ServletContext 关联。然后，在该类中需要注册 Spring MVC 的 DispatcherServlet。最后，在该类中使用 javax.servlet.FilterRegistration.Dynamic 注册字符编码过滤器，防止中文乱码。具体代码如下：

```java
package config;
import javax.servlet.ServletContext;
import javax.servlet.ServletException;
import javax.servlet.ServletRegistration.Dynamic;
import org.springframework.web.WebApplicationInitializer;
import org.springframework.web.context.support.AnnotationConfigWebApplicationContext;
import org.springframework.web.servlet.DispatcherServlet;
public class WebConfig implements WebApplicationInitializer{
    @Override
    public void onStartup(ServletContext arg0) throws ServletException {
        AnnotationConfigWebApplicationContext ctx
        =new AnnotationConfigWebApplicationContext();
        ctx.register(SpringMVCConfig.class);
        //注册 Spring MVC 的 Java 配置类 SpringMVCConfig
        ctx.setServletContext(arg0);//和当前 ServletContext 关联
        /**
         * 注册 Spring MVC 的 DispatcherServlet
         */
        Dynamic servlet =arg0.addServlet("dispatcher", new DispatcherServlet(ctx));
        servlet.addMapping("/");
        servlet.setLoadOnStartup(1);
        /**
         * 注册字符编码过滤器
         */
        javax.servlet.FilterRegistration.Dynamic filter =
                arg0.addFilter("characterEncodingFilter",
                                        CharacterEncodingFilter.class);
        filter.setInitParameter("encoding", "UTF-8");
        filter.addMappingForUrlPatterns(null, false, "/*");
    }
}
```

7. 发布并测试应用

选中 input.jsp 文件名，右击，选择 Run As/Run on Server 发布并测试应用。

4.2.3 实践环节

创建一个 Web 应用 project423pratice,该应用具体实施步骤如下:
(1) 编写一个 JSP 页面 input.jsp,该页面运行效果如图 4.2 所示。
(2) 编写实体类 User。
(3) 编写控制器类,在控制器类中,类型转换器自动将请求过来的值转换成 User 类型。
(4) 编写自定义类型转换器类 UserConverter。
(5) 注册类型转换器。
(6) 编写用户信息输出页面 showUser.jsp,页面效果如图 4.3 所示。

图 4.2 实践环节的首页面　　图 4.3 实践环节的结果页面

4.3 Formatter

Spring MVC 框架的 Formatter<T>与 Converter<S,T>一样,也是一个可以将一种数据类型转换成另一种数据类型的接口。但不同的是,Formatter<T>的源数据类型必须是 String 类型,而 Converter<S,T>的源数据类型是任意数据类型。

在 Web 应用中,由 HTTP 发送的请求数据到控制器中都是以 String 类型获取。因此,在 Web 应用中选择 Formatter<T>比选择 Converter<S,T>更加合理。

4.3.1 内置的格式化转换器

Spring MVC 提供几个内置的格式化转换器,具体如下:
① NumberFormatter:实现 Number 与 String 之间的解析与格式化。
② CurrencyFormatter:实现 Number 与 String 之间的解析与格式化(带货币符号)。
③ PercentFormatter:实现 Number 与 String 之间的解析与格式化(带百分数符号)。
④ DateFormatter:实现 Date 与 String 之间的解析与格式化。

4.3.2 自定义格式化转换器

自定义格式化转换器,就是编写一个实现 org.springframework.format.Formatter 接口的 Java 类。该接口声明如下:

```
public interface Formatter<T>
```

这里的 T 表示由字符串转换的目标数据类型。该接口有 parse 和 print 两个接口方法,自定义格式化转换器类必须重写它们。

```
public T parse(String s, java.util.Locale locale)
public String print(T object, java.util.Locale locale)
```

parse 方法的功能是利用指定的 Locale 将一个 String 类型转换成目标类型,print 方法与之相反,返回目标对象的字符串表示。

下面通过具体应用 ch4_2 讲解自定义格式化转换器的用法。应用 ch4_2 的具体要求如下:

(1) 用户在页面表单中输入信息来创建商品,输入页面效果如图 4.4 所示。

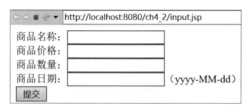

图 4.4　信息输入页面

(2) 控制器使用实体 bean 类 GoodsModel 接收页面提交的请求参数,GoodsModel 类的属性如下:

```
private String goodsname;
private double goodsprice;
private int goodsnumber;
private Date goodsdate;
```

(3) GoodsModel 实体类接收请求参数时,商品名称、价格和数量使用内置的类型转换器完成转换;商品日期需要自定义的格式化转换器完成。

(4) 格式化转换器转换之后的数据显示在 showGoods.jsp 页面,效果如图 4.5 所示。

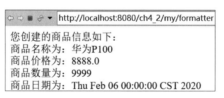

图 4.5　格式化后信息显示页面

由图 4.5 可以看出,日期由字符串值 2020-02-06 格式化成 Date 类型。

欲实现上述应用 ch4_2 的需求,需要完成以下 5 项工作:

① 创建相关视图。
② 创建实体类。
③ 创建控制器类。
④ 创建自定义格式化转换器。
⑤ 注册格式化转换器。

【例4-2】 按照上述步骤,采用自定义格式化转换器完成应用 ch4_2 的需求。具体实现步骤如下。

1. 创建 Maven 项目并添加依赖的 JAR 包

使用 STS 创建一个名为 ch4_2 的 Maven Project,并通过 pom.xml 文件添加项目所依赖的 JAR 包。ch4_2 添加的依赖有 spring-webmvc。pom.xml 文件内容与【例4-1】的相同,不再赘述。

2. 创建相关视图

在 ch4_2 的 src/main/webapp 目录下创建信息输入页面 input.jsp,代码如下:

```jsp
<%@page language="java" contentType="text/html; charset=UTF-8"
    pageEncoding="UTF-8"%>
<%
String path =request.getContextPath();
String basePath = request.getScheme()+"://"+ request.getServerName()+":"+
request.getServerPort()+path+"/";
%>
<!DOCTYPE html>
<html>
<head>
<base href="<%=basePath%>">
<meta charset="UTF-8">
<title>Insert title here</title>
</head>
<body>
    <form action="my/formatter" method="post">
        商品名称:<input type="text" name="goodsname" /><br>商品价格:<input
            type="text" name="goodsprice" /><br>商品数量:<input type="text"
            name="goodsnumber" /><br>商品日期:<input type="text"
            name="goodsdate" />(yyyy-MM-dd)<br><input type="submit"
            value="提交" />
    </form>
</body>
</html>
```

在 ch4_2 的 src/main/webapp/WEB-INF/jsp/目录下创建信息显示页面 showGoods.jsp,代码如下:

```jsp
<%@page language="java" contentType="text/html; charset=UTF-8"
    pageEncoding="UTF-8"%>
<%
String path =request.getContextPath();
String basePath = request.getScheme()+"://"+ request.getServerName()+":"+
```

```
request.getServerPort()+path+"/";
%>
<!DOCTYPE html>
<html>
<head>
<base href="<%=basePath%>">
<meta charset="UTF-8">
<title>Insert title here</title>
</head>
<body>
    您创建的商品信息如下:<br>
    <!--使用 EL 表达式取出 Action 类的属性 goods 的值 -->
    商品名称为:${goods.goodsname }<br>
    商品价格为:${goods.goodsprice }<br>
    商品数量为:${goods.goodsnumber }<br>
    商品日期为:${goods.goodsdate }
</body>
</html>
```

3. 创建实体类

在 ch4_2 的 src/main/java 目录下创建一个名为 pojo 的包,并在该包中创建名为 GoodsModel 的实体类,代码如下:

```
package pojo;
import java.util.Date;
public class GoodsModel {
    private String goodsname;
    private double goodsprice;
    private int goodsnumber;
    private Date goodsdate;
    //省略 setter 和 getter 方法
}
```

4. 创建控制器类

在 ch4_2 的 src/main/java 目录下创建一个名为 controller 的包,并在该包中创建名为 FormatterController 的控制器类,代码如下:

```
package controller;
import org.springframework.stereotype.Controller;
import org.springframework.ui.Model;
import org.springframework.web.bind.annotation.RequestMapping;
import pojo.GoodsModel;
@Controller
```

```
    @RequestMapping("/my")
    public class FormatterController {
        @RequestMapping("/formatter")
        public String myConverter(GoodsModel gm, Model model){
            model.addAttribute("goods",gm);
            return "showGoods";
        }
    }
```

5. 创建格式化转换器类

在 ch4_2 的 src/main/java 目录下创建一个名为 formatter 的包,并在该包中创建名为 MyFormatter 的自定义格式化转换器类。该类是一个组件类,需要使用@Component 注解标注。具体代码如下:

```
package formatter;
import java.text.ParseException;
import java.text.SimpleDateFormat;
import java.util.Date;
import java.util.Locale;
import org.springframework.format.Formatter;
@Component
public class MyFormatter implements Formatter<Date>{
    SimpleDateFormat dateFormat =new SimpleDateFormat("yyyy-MM-dd");
    @Override
    public String print(Date object, Locale arg1) {
        return dateFormat.format(object);
    }
    @Override
    public Date parse(String source, Locale arg1) throws ParseException {
        return dateFormat.parse(source);//Formatter 只能对字符串转换
    }
}
```

6. 创建配置类,并注册格式化转换器

在 ch4_2 应用的 src/main/java 目录中创建一个名为 config 的包,并在该包中创建 Spring MVC 的 Java 配置类 SpringMVCConfig。在该配置类中使用 FormattingConversionService 注册格式化转换器。具体代码如下:

```
package config;
import javax.annotation.PostConstruct;
import org.springframework.beans.factory.annotation.Autowired;
import org.springframework.context.annotation.Bean;
import org.springframework.context.annotation.ComponentScan;
```

```java
import org.springframework.context.annotation.Configuration;
import org.springframework.format.support.FormattingConversionService;
import org.springframework.web.servlet.config.annotation.EnableWebMvc;
import org.springframework.web.servlet.config.annotation.WebMvcConfigurer;
import org.springframework.web.servlet.view.InternalResourceViewResolver;
import formatter.MyFormatter;
@Configuration
@EnableWebMvc    // 开启Spring MVC的支持
@ComponentScan(basePackages = { "controller", "formatter" })    // 扫描基本包
public class SpringMVCConfig implements WebMvcConfigurer {
    @Autowired
    private FormattingConversionService formatConversionService;//用于注册格
                                                                //式化转换器

    @Autowired
    private MyFormatter myFormatter;
    /**
     * 配置视图解析器
     */
    @Bean
    public InternalResourceViewResolver getViewResolver() {
        InternalResourceViewResolver viewResolver =new
                                         InternalResourceViewResolver();
        viewResolver.setPrefix("/WEB-INF/jsp/");
        viewResolver.setSuffix(".jsp");
        return viewResolver;
    }
    /**
     * 注册格式化转换器
     */
    @PostConstruct
    /**
     * 被@PostConstruct修饰的方法会在服务器加载Servlet的时候运行,并且只会被服务
       器调用一次,类似于Servlet的inti()方法。
     * 被@PostConstruct修饰的方法会在构造函数之后,init()方法之前运行。
     */
    public void initEditableAvlidation() {
        if (formatConversionService !=null) {
            formatConversionService.addFormatter(myFormatter);
               //可以添加多个格式化转换器
        }
    }
}
```

在ch4_2应用的config包中创建Web的Java类WebConfig。该类与【例4-1】的相

同,不再赘述。

7. 发布并测试应用

选中 input.jsp 文件名,右击,选择 Run As/Run on Server 发布并测试应用。

4.3.3 实践环节

创建一个 Web 应用 project433pratice,在该应用中将 4.2.3 节的自定义类型转换器修改成自定义格式化转换器。

4.4 本章小结

本章重点讲解了自定义类型转换器和格式化转换器的实现和注册。但在实际应用中,开发者很少自定义类型转换器和格式化类型转换器,一般都是使用内置的转换器。

习 题 4

1. 在 MVC 框架中,为什么要进行类型转换?
2. Converter 与 Formatter 的区别是什么?
3. 在 Spring MVC 框架中,如何自定义类型转换器类,又如何注册类型转换器?
4. 在 Spring MVC 框架中,如何自定义格式化转换器类,又如何注册格式化转换器?

数据绑定和表单标签库

学习目的与要求

本章主要讲解数据绑定和表单标签库。通过本章的学习,理解数据绑定的基本原理,掌握表单标签库的用法。

本章主要内容

- 数据绑定。
- 表单标签库。
- 数据绑定应用。
- JSON 数据交互。

数据绑定是将用户参数输入值绑定到领域模型的一种特性。在 Spring MVC 的 Controller 和 View 参数数据传递中,所有 HTTP 请求参数的类型均为字符串,如果模型需要绑定的类型为 double 或 int,则需要手动进行类型转换,而有了数据绑定后,就不再需要手动将 HTTP 请求中的 String 类型转换为模型需要的类型。数据绑定的另一个好处是,当输入验证失败时,会重新生成一个 HTML 表单,无须重新填写输入字段。

在 Spring MVC 中,为了方便、高效地使用数据绑定,还需要学习 Spring 的表单标签库。

5.1 数据绑定

在 Spring MVC 框架中,数据绑定有这样几层含义:绑定请求参数输入值到领域模型(如 3.2 节)、模型数据到视图的绑定(输入验证失败时)、模型数据到表单元素的绑定(如下拉列表选项值由控制器初始化)。有关数据绑定的示例,请参见 5.3 节的数据绑定应用。

5.2 Spring 的表单标签库

表单标签库中包含了可以用在 JSP 页面中渲染 HTML 元素的标签。JSP 页面使用 Spring 表单标签库时,必须在 JSP 页面开头声明 taglib 指令,指令代码如下:

```
<%@taglib prefix="form" uri="http://www.springframework.org/tags/form" %>
```

表单标签库中有 form、input、password、hidden、textarea、checkbox、checkboxes、radiobutton、radiobuttons、select、option、options、errors 等标签。

form:渲染表单元素。

input:渲染<input type="text"/>元素。

password:渲染<input type="password"/>元素。

hidden:渲染<input type="hidden"/>元素。

textarea:渲染 textarea 元素。

checkbox:渲染一个<input type="checkbox"/>元素。

checkboxes:渲染多个<input type="checkbox"/>元素。

radiobutton:渲染一个<input type="radio"/>元素。

radiobuttons:渲染多个<input type="radio"/>元素。

select:渲染一个选择元素。

option:渲染一个选项元素。

options:渲染多个选项元素。

errors:在 span 元素中渲染字段错误。

5.2.1 表单标签

表单标签的语法格式如下:

```
<form:form modelAttribute="xxx" method="post" action="xxx">
    ...
</form:form>
```

除了具有 HTML 表单元素属性外,表单标签还有具有 acceptCharset、cssClass、cssStyle、htmlEscape 和 modelAttribute 等属性。各属性含义如下。

acceptCharset:定义服务器接受的字符编码列表。

cssClass：定义应用到 form 元素的 CSS 类。
cssStyle：定义应用到 form 元素的 CSS 样式。
htmlEscape：true 或 false，表示是否进行 HTML 转义。
modelAttribute：暴露 form backing object 的模型属性名称，缺省为 command。

其中，modelAttribute 属性值绑定一个 JavaBean 对象。假设控制器类 UserController 的方法 inputUser，是返回 userAdd.jsp 的请求处理方法。

inputUser 方法的代码如下：

```
@RequestMapping(value ="/input")
public String inputUser(Model model) {
    ...
    model.addAttribute("user", new User());
    return "userAdd";
}
```

userAdd.jsp 的表单标签代码如下：

```
<form:form modelAttribute="user"  method="post" action="user/save">
    ...
</form:form>
```

注意：在 inputUser 方法中，如果没有 Model 属性 user，userAdd.jsp 页面就会抛出异常，因为表单标签无法找到在其 modelAttribute 属性中指定的 form backing object。

5.2.2　input 标签

input 标签，语法格式如下：

```
<form:input path="xxx"/>
```

该标签除了 cssClass、cssStyle、htmlEscape 属性外，还有一个最重要的属性 path。path 属性将文本框输入值绑定到 form backing object 的一个属性。示例代码如下：

```
<form:form modelAttribute="user"  method="post" action="user/save">
    <form:input path="userName"/>
</form:form>
```

上述代码将输入值绑定到 user 对象的 userName 属性。

5.2.3　password 标签

password 标签的语法格式如下：

```
<form:password path="xxx"/>
```

该标签与 input 标签用法完全一致，不再赘述。

5.2.4　hidden 标签

hidden 标签的语法格式如下：

```
<form:hidden path="xxx"/>
```

该标签与 input 标签用法基本一致,只不过它不可显示,不支持 cssClass 和 cssStyle 属性。

5.2.5 textarea 标签

textarea 基本就是一个支持多行输入的 input 元素,语法格式如下:

```
<form:textarea path="xxx"/>
```

该标签与 input 标签用法完全一致,不再赘述。

5.2.6 checkbox 标签

checkbox 标签的语法格式如下:

```
<form:checkbox path="xxx" value="xxx"/>
```

多个 path 相同的 checkbox 标签是一个选项组,允许多选。选项值绑定到一个数组属性。示例代码如下:

```
<form:checkbox path="friends" value="张三"/>张三
<form:checkbox path="friends" value="李四"/>李四
<form:checkbox path="friends" value="王五"/>王五
<form:checkbox path="friends" value="赵六"/>赵六
```

上述示例代码中复选框的值绑定到一个字符串数组属性 friends(String[] friends)。该标签的其他用法与 input 标签基本一致,不再赘述

5.2.7 checkboxes 标签

checkboxes 标签渲染多个复选框,是一个选项组,等价于多个 path 相同的 checkbox 标签。它有 3 个非常重要的属性:items、itemLabel 和 itemValue。

items:用于生成 input 元素的 Collection、Map 或 Array。

itemLabel:items 属性中指定的集合对象的属性,为每个 input 元素提供 label。

itemValue:items 属性中指定的集合对象的属性,为每个 input 元素提供 value。

checkboxes 标签语法格式如下:

```
<form:checkboxes items="xxx" path="xxx"/>
```

示例代码如下:

```
<form:checkboxes items="${hobbys}" path="hobby" />
```

上述示例代码是将 model 属性 hobbys 的内容(集合元素)渲染为复选框。在 itemLabel 和 itemValue 缺省的情况下,如果集合是数组,复选框的 label 和 value 相同;如果是 Map 集合,复选框的 label 是 Map 的值(value),复选框的 value 是 Map 的关键字(key)。

5.2.8　radiobutton 标签

radiobutton 标签的语法格式如下：

```
<form:radiobutton path="xxx" value="xxx"/>
```

多个 path 相同的 radiobutton 标签是一个选项组，只允许单选。

5.2.9　radiobuttons 标签

radiobuttons 标签渲染多个 radio，是一个选项组，等价于多个 path 相同的 radiobutton 标签。radiobuttons 标签的语法格式如下：

```
<form:radiobuttons path="xxx" items="xxx"/>
```

该标签的 itemLabel 和 itemValue 属性与 checkboxes 标签的 itemLabel 和 itemValue 属性完全一样，但只允许单选。

5.2.10　select 标签

select 标签的选项可来自其属性 items 指定的集合，或来自一个嵌套的 option 标签，或来自 options 标签。语法格式如下：

```
<form:select path="xxx" items="xxx" />
```

或

```
<form:select path="xxx" items="xxx" >
    <option value="xxx">xxx</option>
</ form:select>
```

或

```
<form:select path="xxx">
    <form:options items="xxx"/>
</form:select>
```

该标签的 itemLabel 和 itemValue 属性与 checkboxes 标签的 itemLabel 和 itemValue 属性完全一样。

5.2.11　options 标签

options 标签生成一个 select 标签的选项列表。因此，需要与 select 标签一同使用，具体用法参见 5.2.10 节中的 select 标签。

5.2.12　errors 标签

errors 标签渲染一个或多个 span 元素，每个 span 元素包含一个错误消息。它可以用于显示一个特定的错误消息，也可以显示所有错误消息。语法格式如下：

```
<form:errors path="*"/>
```

或

```
<form:errors path="xxx"/>
```

其中,"*"表示显示所有错误消息;"xxx"表示显示由"xxx"指定的特定错误消息。

5.3 数据绑定应用

为了进一步讲解数据绑定和表单标签,本节给出了一个应用范例 ch5_1。

【例 5-1】 ch5_1 应用中实现了 User 类属性和 JSP 页面中表单参数的绑定,同时 JSP 页面中分别展示了 input、textarea、checkbox、checkboxs、select 等标签。具体实现步骤如下。

5.3.1 创建 Maven 项目并添加相关依赖

在 ch5_1 应用中需要使用 JSTL,因此,不仅需要将 Spring MVC 相关依赖添加到 pom.xml 文件中,还需要将 JSTL 相关依赖添加到 pom.xml 文件中。pom.xml 文件内容如下:

```
<project xmlns="http://maven.apache.org/POM/4.0.0"
    xmlns:xsi="http://www.w3.org/2001/XMLSchema-instance"
    xsi:schemaLocation="http://maven.apache.org/POM/4.0.0
https://maven.apache.org/xsd/maven-4.0.0.xsd">
    <modelVersion>4.0.0</modelVersion>
    <groupId>com.cn.chenheng</groupId>
    <artifactId>ch5_1</artifactId>
    <version>0.0.1-SNAPSHOT</version>
    <packaging>war</packaging>
    <properties>
        <!--spring 版本号 -->
        <spring.version>5.2.3.RELEASE</spring.version>
    </properties>
    <dependencies>
        <dependency>
            <groupId>org.springframework</groupId>
            <artifactId>spring-webmvc</artifactId>
            <version>${spring.version}</version>
        </dependency>
        <!--JSTL 依赖 -->
        <dependency>
            <groupId>javax.servlet</groupId>
            <artifactId>jstl</artifactId>
```

```
            <version>1.2</version>
        </dependency>
    </dependencies>
</project>
```

5.3.2　Spring MVC 及 Web 相关配置

1. Spring MVC 配置类

在应用 ch5_1 的 src/main/java 目录中创建一个名为 config 的包，并在该包中创建 Spring MVC 的配置类 SpringMVCConfig。SpringMVCConfig 的代码如下：

```
package config;
import org.springframework.context.annotation.Bean;
import org.springframework.context.annotation.ComponentScan;
import org.springframework.context.annotation.Configuration;
import org.springframework.web.servlet.config.annotation.EnableWebMvc;
import org.springframework.web.servlet.config.annotation.WebMvcConfigurer;
import org.springframework.web.servlet.view.InternalResourceViewResolver;
@Configuration
@EnableWebMvc // 开启 spring mvc 的支持
@ComponentScan(basePackages = { "controller", "service"}) // 扫描基本包
public class SpringMVCConfig implements WebMvcConfigurer {
    /**
     * 配置视图解析器
     */
    @Bean
    public InternalResourceViewResolver getViewResolver() {
        InternalResourceViewResolver viewResolver = new
        InternalResourceViewResolver();
        viewResolver.setPrefix("/WEB-INF/jsp/");
        viewResolver.setSuffix(".jsp");
        return viewResolver;
    }
}
```

2. Web 配置类

在应用 ch5_1 的 config 包中创建 Web 的配置类 WebConfig。为了避免中文乱码问题，需要在配置类 WebConfig 中添加编码过滤器，同时将 JSP 页面编码设置为 UTF-8，form 表单的提交方式必须为 post。WebConfig 的代码如下：

```
package config;
import javax.servlet.ServletContext;
import javax.servlet.ServletException;
```

```java
import javax.servlet.ServletRegistration.Dynamic;
import org.springframework.web.WebApplicationInitializer;
import org.springframework.web.context.support.
AnnotationConfigWebApplicationContext;
import org.springframework.web.filter.CharacterEncodingFilter;
import org.springframework.web.servlet.DispatcherServlet;
public class WebConfig implements WebApplicationInitializer{
    @Override
    public void onStartup(ServletContext arg0) throws ServletException {
        AnnotationConfigWebApplicationContext ctx
        =new AnnotationConfigWebApplicationContext();
        ctx.register(SpringMVCConfig.class);//注册 Spring MVC 的 Java 配置
                                            //类 SpringMVCConfig
        ctx.setServletContext(arg0);//和当前 ServletContext 关联
        /**
         * 注册 Spring MVC 的 DispatcherServlet
         */
        Dynamic servlet =arg0.addServlet("dispatcher", new DispatcherServlet(ctx));
        servlet.addMapping("/");
        servlet.setLoadOnStartup(1);
        /**
         * 注册字符编码过滤器
         */
        javax.servlet.FilterRegistration.Dynamic filter =
                arg0.addFilter("characterEncodingFilter",
                CharacterEncodingFilter.class);
        filter.setInitParameter("encoding", "UTF-8");
        filter.addMappingForUrlPatterns(null, false, "/*");
    }
}
```

5.3.3 领域模型

应用中实现了 User 类属性和 JSP 页面中表单参数的绑定，User 类包含了和表单参数名对应的属性，以及属性的 set 和 get 方法。在 ch5_1 应用的 src/main/java 目录下创建一个名为 pojo 的包，并在该包中创建 User 类。User 类的代码如下：

```java
package pojo;
public class User {
    private String userName;
    private String[] hobby;//兴趣爱好
    private String[] friends;//朋友
    private String carrer;
    private String houseRegister;
```

```
    private String remark;
    //省略 setter 和 getter 方法
}
```

5.3.4　Service 层

应用中使用了 Service 层,在 Service 层使用静态集合变量 users 模拟数据库存储用户信息,包括添加用户和查询用户两个功能方法。在 ch5_1 应用的 src/main/java 目录下创建一个名为 service 的包,并在该包中创建 UserService 接口和 UserServiceImpl 实现类。

UserService 接口的代码如下:

```
package service;
import java.util.ArrayList;
import pojo.User;
public interface UserService {
    boolean addUser(User u);
    ArrayList<User>getUsers();
}
```

UserServiceImpl 实现类的代码如下:

```
package service;
import java.util.ArrayList;
import org.springframework.stereotype.Service;
import pojo.User;
@Service
public class UserServiceImpl implements UserService{
    //使用静态集合变量 users 模拟数据库
    private static ArrayList<User>users =new ArrayList<User>();
    @Override
    public boolean addUser(User u) {
        if(!"IT 民工".equals(u.getCarrer())){//不允许添加 IT 民工
            users.add(u);
            return true;
        }
        return false;
    }
    @Override
    public ArrayList<User>getUsers() {
        return users;
    }
}
```

5.3.5　Controller 层

在 Controller 类 UserController 中定义了请求处理方法,其中包括处理 user/input

请求的 inputUser 方法,以及 user/save 请求的 addUser 方法,其中在 addUser 方法中用到了重定向。在 UserController 类中,通过@Autowired 注解在 UserController 对象中主动注入 UserService 对象,实现对 user 对象的添加和查询等操作;通过 model 的 addAttribute 方法将 User 类对象、HashMap 类型的 hobbys 对象、String[]类型的 carrers 对象以及 String[]类型的 houseRegisters 对象传递给 View(userAdd.jsp)。在 ch5_1 应用的 src/main/java 目录下创建一个名为 controller 的包,并在该包中创建 UserController 控制器类。UserController 类的代码如下:

```java
package controller;
import java.util.HashMap;
import java.util.List;
import org.apache.commons.logging.Log;
import org.apache.commons.logging.LogFactory;
import org.springframework.beans.factory.annotation.Autowired;
import org.springframework.stereotype.Controller;
import org.springframework.ui.Model;
import org.springframework.web.bind.annotation.ModelAttribute;
import org.springframework.web.bind.annotation.RequestMapping;
import pojo.User;
import service.UserService;
@Controller
@RequestMapping("/user")
public class UserController {
    // 得到一个用来记录日志的对象,这样打印信息的时候能够标记打印的是那个类的信息
    private static final Log logger =LogFactory.getLog(UserController.class);
    @Autowired
    private UserService userService;
    @RequestMapping(value ="/input")
    public String inputUser(Model model) {
        HashMap<String, String>hobbys =new HashMap<String, String>();
        hobbys.put("篮球", "篮球");
        hobbys.put("乒乓球", "乒乓球");
        hobbys.put("电玩", "电玩");
        hobbys.put("游泳", "游泳");
         // 如果 model 中没有 user 属性,userAdd.jsp 会抛出异常,因为表单标签无法找到
        // modelAttribute 属性指定的 form backing object
        model.addAttribute("user", new User());
        model.addAttribute("hobbys", hobbys);
        model.addAttribute("carrers", new String[] { "教师", "学生", "coding 搬运工", "IT 民工", "其他" });
        model.addAttribute("houseRegisters", new String[] { "北京", "上海", "广州", "深圳", "其他" });
        return "userAdd";
```

```java
        }
        @RequestMapping(value = "/save")
        public String addUser(@ModelAttribute User user, Model model) {
            if (userService.addUser(user)) {
                logger.info("成功");
                return "redirect:/user/list";
            } else {
                logger.info("失败");
                HashMap<String, String>hobbys =new HashMap<String, String>();
                hobbys.put("篮球", "篮球");
                hobbys.put("乒乓球", "乒乓球");
                hobbys.put("电玩", "电玩");
                hobbys.put("游泳", "游泳");
                // 这里不需要 model.addAttribute("user", new
                // User()),因为@ModelAttribute 指定 form backing object
                model.addAttribute("hobbys", hobbys);
                model.addAttribute("carrers", new String[] { "教师", "学生", "coding
                                    搬运工", "IT民工", "其他" });
                model.addAttribute("houseRegisters", new String[] { "北京", "上海",
                                    "广州", "深圳", "其他" });
                return "userAdd";
            }
        }
        @RequestMapping(value = "/list")
        public String listUsers(Model model) {
            List<User>users =userService.getUsers();
            model.addAttribute("users", users);
            return "userList";
        }
    }
```

5.3.6 View 层

View 层包含两个 JSP 页面,一个是信息输入页面 userAdd.jsp,一个是信息显示页面 userList.jsp。在 ch5_1 应用的 src/main/webapp/WEB-INF/jsp/ 目录下创建此两个 JSP 页面。

在 userAdd.jsp 页面中,将 Map 类型的 hobbys 绑定到 checkboxes 上,将 String[] 类型的 carrers 和 houseRegisters 绑定到 select 上,实现通过 option 标签对 select 添加选项,同时表单的 method 方法需指定为 post,以避免中文乱码问题。

在 userList.jsp 页面中使用 JSTL 标签遍历集合中的用户信息,JSTL 的相关知识参见本书的相关内容。

userAdd.jsp 的代码如下:

```jsp
<%@page language="java" contentType="text/html; charset=UTF-8" pageEncoding="UTF-8"%>
<%@taglib prefix="form" uri="http://www.springframework.org/tags/form" %>
<%
String path =request.getContextPath();
String basePath = request.getScheme()+"://"+request.getServerName()+":"+request.getServerPort()+path+"/";
%>
<!DOCTYPE html>
<html>
<head>
<base href="<%=basePath%>">
<meta charset="UTF-8">
<title>Insert title here</title>
</head>
<body>
<form:form modelAttribute="user" method="post" action="user/save">
    <fieldset>
        <legend>添加一个用户</legend>
        <p>
            <label>用户名:</label>
            <form:input path="userName"/>
        </p>
        <p>
            <label>爱好:</label>
            <form:checkboxes items="${hobbys}" path="hobby" />
        </p>
        <p>
            <label>朋友:</label>
            <form:checkbox path="friends" value="张三"/>张三
            <form:checkbox path="friends" value="李四"/>李四
            <form:checkbox path="friends" value="王五"/>王五
            <form:checkbox path="friends" value="赵六"/>赵六
        </p>
        <p>
            <label>职业:</label>
            <form:select path="carrer">
                <option/>请选择职业
                <form:options items="${carrers}"/>
            </form:select>
        </p>
        <p>
            <label>户籍:</label>
            <form:select path="houseRegister">
```

```
            <option/>请选择户籍
            <form:options items="${houseRegisters }"/>
        </form:select>
    </p>
    <p>
        <label>个人描述:</label>
        <form:textarea path="remark" rows="5"/>
    </p>
    <p id="buttons">
        <input id="reset" type="reset">
        <input id="submit" type="submit" value="添加">
    </p>
    </fieldset>
</form:form>
</body>
</html>
```

userList.jsp 的代码如下:

```
<%@page language="java" contentType="text/html; charset=UTF-8" pageEncoding="UTF-8"%>
<%@taglib uri="http://java.sun.com/jsp/jstl/core" prefix="c" %>
<%
String path =request.getContextPath();
String basePath = request.getScheme() +"://"+ request.getServerName() +":"+ request.getServerPort()+path+"/";
%>
<!DOCTYPE html>
<html>
<head>
<base href="<%=basePath%>">
<meta charset="UTF-8">
<title>Insert title here</title>
</head>
<body>
    <h1>用户列表</h1>
    <a href="<c:url value="user/input"/>">继续添加</a>
    <table>
        <tr>
            <th>用户名</th>
            <th>兴趣爱好</th>
            <th>朋友</th>
            <th>职业</th>
            <th>户籍</th>
            <th>个人描述</th>
```

```
            </tr>
            <!--JSTL 标签,请参考本书的相关内容 -->
            <c:forEach items="${users}" var="user">
                <tr>
                    <td>${user.userName }</td>
                    <td>
                        <c:forEach items="${user.hobby }" var="hobby">
                            ${hobby } 
                        </c:forEach>
                    </td>
                    <td>
                        <c:forEach items="${user.friends }" var="friend">
                            ${friend } 
                        </c:forEach>
                    </td>
                    <td>${user.carrer }</td>
                    <td>${user.houseRegister }</td>
                    <td>${user.remark }</td>
                </tr>
            </c:forEach>
        </table>
    </body>
</html>
```

5.3.7 测试应用

发布应用后,通过地址 http://localhost:8080/ch5_1/user/input 测试应用,添加用户信息页面效果如图 5.1 所示。

如果在图 5.1 的职业中选择"IT 民工",添加失败。失败后还回到添加页面,输入过的信息不再输入,自动回填(必须结合 form 标签)。自动回填是数据绑定的一个优点。失败页面如图 5.2 所示。

图 5.1 添加用户信息页面

图 5.2 添加用户信息失败页面

在图 5.1 中输入正确信息，添加成功后，重定向到信息显示页面，效果如图 5.3 所示。

图 5.3 信息显示页面

5.4 实践环节

参考应用 ch5_1 创建应用 practice55。在应用 practice55 中创建两个视图页面 addGoods.jsp 和 goodsList.jsp。addGoods.jsp 页面的显示效果如图 5.4 所示，goodsList.jsp 页面的显示效果如图 5.5 所示。

图 5.4 添加商品页面　　图 5.5 商品显示页面

具体要求如下：

（1）商品类型由控制器类 GoodsController 的方法 inputGoods 进行初始化。GoodsController 类中共有 3 个方法：inputGoods、addGoods 和 listGoods。

（2）使用 Goods 模型类封装请求参数。

（3）使用 Service 层，在 Service 的实现类中使用静态集合变量模拟数据库存储商品信息，在控制器类中使用@Autowired 注解 Service。

（4）通过地址 http://localhost:8080/practice55/goods/input 访问 addGoods.jsp 页面。

（5）其他注意事项参见应用 ch5_1。

5.5 JSON 数据交互

在数据绑定的过程中，Spring MVC 需要对传递数据的格式和类型进行转换，它既可以转换 String 等类型的数据，也可以转换 JSON 等其他类型的数据。本节将针对 Spring MVC 中 JSON 类型的数据交互进行讲解。

5.5.1 JSON 概述

JS 对象标记(JavaScript Object Notation，JSON)是一种轻量级的数据交换格式。与

XML 一样,JSON 也是基于纯文本的数据格式。它有两种数据结构。

1. 对象结构

对象结构以"{"开始,以"}"结束。中间部分由 0 个或多个以英文","分隔的 key/value 对构成,key 和 value 之间以英文":"分隔。对象结构的语法结构如下:

```
{
    key1:value1,
    key2:value2,
    ...
}
```

其中,key 必须为 String 类型,value 可以是 String、Number、Object、Array 等数据类型。例如,一个 person 对象包含姓名、密码、年龄等信息,使用 JSON 的表示形式如下:

```
{
    "pname":"陈恒",
    "password":"123456",
    "page":40
}
```

2. 数组结构

数组结构以"["开始,以"]"结束。中间部分由 0 个或多个以英文","分隔的值的列表组成。数组结构的语法结构如下:

```
[
    value1,
    value2,
    ...
]
```

上述两种(对象、数组)数据结构也可以分别组合构成更为复杂的数据结构。例如:一个 student 对象包含 sno、sname、hobby 和 college 对象,其 JSON 的表示形式如下:

```
{
    "sno":"201802228888",
    "sname":"陈恒",
    "hobby":["篮球","足球"],
    "college":{
        "cname":"清华大学",
        "city":"北京"
    }
}
```

5.5.2　JSON 数据转换

为实现浏览器与控制器类之间的 JSON 数据交互，Spring MVC 提供了 MappingJackson2HttpMessageConverter 实现类默认处理 JSON 格式请求响应。该实现类利用 Jackson 开源包读写 JSON 数据，将 Java 对象转换为 JSON 对象和 XML 文档，同时也可以将 JSON 对象和 XML 文档转换为 Java 对象。

Jackson 开源包及其描述如下。

- jackson-annotations.jar：JSON 转换的注解包。
- jackson-core.jar：JSON 转换的核心包。
- jackson-databind.jar：JSON 转换的数据绑定包。

以上 3 个 Jackson 开源包的最新版本是 2.10.2，读者可通过地址 http://mvnrepository.com/artifact/com.fasterxml.jackson.core 下载。

使用注解开发时，需要用到两个重要的 JSON 格式转换注解，分别是@RequestBody 和@ResponseBody。

- @RequestBody：用于将请求体中的数据绑定到方法的形参中，该注解应用在方法的形参上。
- @ResponseBody：用于直接返回 return 对象，该注解应用在方法上。

下面通过一个案例演示如何进行 JSON 数据交互。

【例 5-2】　JSON 数据交互。

具体实现步骤如下。

1. 创建 Maven 项目并添加相关依赖

创建 Maven 项目 ch5_2，在 ch5_2 中需要使用 Spring MVC 框架进行 JSON 数据交互。因此，不仅需要将 spring-webmvc 依赖添加到 pom.xml 文件中，还需要将 JSON 相关依赖添加到 pom.xml 文件中。因为 jackson-databind 依赖于 jackson-core 和 jackson-annotations，所以，这里只需添加 spring-webmvc 和 jackson-databind 依赖。pom.xml 文件内容如下：

```
<project xmlns="http://maven.apache.org/POM/4.0.0"
    xmlns:xsi="http://www.w3.org/2001/XMLSchema-instance"
    xsi:schemaLocation="http://maven.apache.org/POM/4.0.0
https://maven.apache.org/xsd/maven-4.0.0.xsd">
    <modelVersion>4.0.0</modelVersion>
    <groupId>com.cn.chenheng</groupId>
    <artifactId>ch5_2</artifactId>
    <version>0.0.1-SNAPSHOT</version>
    <packaging>war</packaging>
    <properties>
        <!--spring 版本号 -->
        <spring.version>5.2.3.RELEASE</spring.version>
```

```xml
        </properties>
        <dependencies>
            <dependency>
                <groupId>org.springframework</groupId>
                <artifactId>spring-webmvc</artifactId>
                <version>${spring.version}</version>
            </dependency>
            <dependency>
                <groupId>com.fasterxml.jackson.core</groupId>
                <artifactId>jackson-databind</artifactId>
                <version>2.10.2</version>
            </dependency>
        </dependencies>
</project>
```

2. Spring MVC 及 Web 相关配置

在 Spring MVC 配置类中，除了配置视图解析器，还需要配置不需要 DispatcherServlet 转发的 BootStrap 等静态资源。Spring MVC 配置类 SpringMVCConfig 的代码如下：

```java
package config;
import org.springframework.context.annotation.Bean;
import org.springframework.context.annotation.ComponentScan;
import org.springframework.context.annotation.Configuration;
import org.springframework.web.servlet.config.annotation.EnableWebMvc;
import org.springframework.web.servlet.config.annotation.ResourceHandlerRegistry;
import org.springframework.web.servlet.config.annotation.WebMvcConfigurer;
import org.springframework.web.servlet.view.InternalResourceViewResolver;
@Configuration
@EnableWebMvc // 开启 spring mvc 的支持
@ComponentScan(basePackages = { "controller"}) // 扫描基本包
public class SpringMVCConfig implements WebMvcConfigurer {
    /**
     * 配置视图解析器
     */
    @Bean
    public InternalResourceViewResolver getViewResolver() {
        InternalResourceViewResolver viewResolver = new InternalResourceViewResolver();
        viewResolver.setPrefix("/WEB-INF/jsp/");
        viewResolver.setSuffix(".jsp");
        return viewResolver;
    }
    /**
```

```
 * 配置静态资源(不需要 DispatcherServlet 转发的请求)
 */
@Override
public void addResourceHandlers(ResourceHandlerRegistry registry) {
    registry.addResourceHandler("/static/**").addResourceLocations
            ("/static/");
    }
}
```

Web 配置与【例 5-1】相同，不再赘述。

3. 创建 POJO 类

在 ch5_2 的 src/main/java 目录下创建一个名为 pojo 的包，并在包中创建 POJO 类 Person，代码如下：

```
package pojo;
public class Person {
    private String pname;
    private String password;
    private Integer page;
    //省略 setter 和 getter 方法
}
```

4. 创建 JSP 页面测试 JSON 数据交互

在 ch5_2 的 src/main/webapp 目录下创建页面 index.jsp 来测试 JSON 数据交互。在 index.jsp 页面中使用 BootStrap 美化页面，并编写一个测试 JSON 交互的表单，当单击"测试"按钮时，执行页面中的 testJson() 函数。在该函数中使用 jQuery 的 AJAX 方式，将 JSON 格式的数据传递给/testJson 结尾的请求中。index.jsp 的代码如下：

```
<%@page language="java" contentType="text/html; charset=UTF-8" pageEncoding=
"UTF-8"%>
<%
String path = request.getContextPath();
String basePath = request.getScheme()+"://"+request.getServerName()+":"+
request.getServerPort()+path+"/";
%>
<!DOCTYPE html>
<html>
<head>
<base href="<%=basePath%>">
<meta charset="UTF-8">
<title>Insert title here</title>
<link rel="stylesheet" href="static/css/bootstrap.min.css" />
```

```html
<script type="text/javascript" src="static/js/jquery.min.js"></script>
<script type="text/javascript">
function testJson() {
    //获取输入的值 pname 为 id
    var pname = $("#pname").val();
    var password = $("#password").val();
    var page = $("#page").val();
    $.ajax({
        //发送请求的 URL 字符串
        url : "testJson",
        //请求类型
        type : "post",
        //定义发送请求的数据格式为 JSON 字符串
        contentType : "application/json",
        //data 表示发送的数据
        data : JSON.stringify({pname:pname,password:password,page:page}),
        //成功响应的结果
        success : function(data){
            if(data !=null){
                //返回一个 Person 对象
                //alert("输入的用户名:" +data.pname +",密码:" +data.password +
                //",年龄:" +data.page);
                //ArrayList<Person>对象
                /**for(var i =0; i <data.length; i++){
                    alert(data[i].pname);
                }**/
                //返回一个 Map<String, Object>对象
                //alert(data.pname);//pname 为 key
                //返回一个 List<Map<String, Object>>对象
                for(var i =0; i <data.length; i++){
                    alert(data[i].pname);
                }
            }
        },
        //请求出错
        error:function(){
            alert("数据发送失败");
        }
    });
}
</script>
</head>
<body>
    <div class="panel panel-primary">
```

第 5 章　数据绑定和表单标签库

```html
            <div class="panel-heading">
                <h3 class="panel-title">处理 JSON 数据</h3>
            </div>
        </div>
<div class="container">
    <div>
        <h4>添加用户</h4>
    </div>
    <div class="row">
        <div class="col-md-6 col-sm-6">
            <form class="form-horizontal" action="">
                <div class="form-group">
                    <div class="input-group col-md-6">
                        <span class="input-group-addon">
                            <i class="glyphicon glyphicon-pencil"></i>
                        </span>
                        <input class="form-control" type="text"
                         id="pname" placeholder="请输入用户名"/>
                    </div>
                </div>
                <div class="form-group">
                    <div class="input-group col-md-6">
                        <span class="input-group-addon">
                            <i class="glyphicon glyphicon-pencil"></i>
                        </span>
                        <input class="form-control" type="password"
                         id="password" placeholder="请输入密码"/>
                    </div>
                </div>
                <div class="form-group">
                    <div class="input-group col-md-6">
                        <span class="input-group-addon">
                            <i class="glyphicon glyphicon-pencil"></i>
                        </span>
                        <input class="form-control" type="text"
                         id="page" placeholder="请输入年龄"/>
                    </div>
                </div>
                <div class="form-group">
                    <div class="col-md-6">
                        <div class="btn-group btn-group-justified">
                            <div class="btn-group">
                                <button type="button" onclick="testJson()" class="btn
                                btn-success">
```

```
                    <span class="glyphicon glyphicon-share">
                    </span>
                     测试
                </button>
              </div>
            </div>
          </div>
        </form>
      </div>
    </div>
  </div>
</body>
</body>
</html>
```

从本节开始,书中后续的 JSP 页面尽量使用 BootStrap 美化,BootStrap 的知识不属于本书的范畴,请读者参考相关内容学习。另外,与 BootStrap 相关的 CSS 与 JS 代码不在书中提供,请参考本书提供的配套源代码。

5. 创建控制器类

在 ch5_2 的 src/main/java 目录下创建一个名为 controller 的包,并在该包中创建一个用于用户操作的控制器类 TestController,代码如下:

```
package controller;
import java.util.ArrayList;
import java.util.HashMap;
import java.util.List;
import java.util.Map;
import org.springframework.stereotype.Controller;
import org.springframework.web.bind.annotation.RequestBody;
import org.springframework.web.bind.annotation.RequestMapping;
import org.springframework.web.bind.annotation.ResponseBody;
import pojo.Person;
@Controller
public class TestController {
    /**
     * 接收页面请求的 JSON 数据,并返回 JSON 格式结果
     */
    @RequestMapping("/testJson")
    @ResponseBody
    public List<Map<String, Object>>testJson(@RequestBody Person user) {
        //打印接收的 JSON 格式数据
        System.out.println("pname=" +user.getPname() +", password=" +
```

```java
            user.getPassword() +",page=" +user.getPage());
    //返回 Person 对象
    //return user;
    /**ArrayList<Person>allp =new ArrayList<Person>();
    Person p1 =new Person();
    p1.setPname("陈恒 1");
    p1.setPassword("123456");
    p1.setPage(80);
    allp.add(p1);
    Person p2 =new Person();
    p2.setPname("陈恒 2");
    p2.setPassword("78910");
    p2.setPage(90);
    allp.add(p2);
    //返回 ArrayList<Person>对象
    return allp;
    **/
    Map<String, Object>map =new HashMap<String, Object>();
    map.put("pname", "陈恒 2");
    map.put("password", "123456");
    map.put("page", 25);
    //返回一个 Map<String, Object>对象
    //return map;
    //返回一个 List<Map<String, Object>>对象
    List<Map<String, Object>>allp =new ArrayList<Map<String, Object>>();
    allp.add(map);
    Map<String, Object>map1 =new HashMap<String, Object>();
    map1.put("pname", "陈恒 3");
    map1.put("password", "54321");
    map1.put("page", 55);
    allp.add(map1);
    return allp;
    }
}
```

在上述控制器类中编写了接收和响应 JSON 格式数据的 testJson()方法。方法中的 @RequestBody 注解用于将前端请求体中的 JSON 格式数据绑定到形参 user 上，@ResponseBody 注解用于直接返回对象。

6. 运行 index.jsp 页面，测试程序

将 ch5_2 应用发布到 Tomcat 服务器并启动服务器，在浏览器中访问地址 http://localhost:8080/ch5_2/index.jsp，运行效果如图 5.6 所示。

图 5.6　index.jsp 测试页面

5.6　本章小结

本章首先介绍了 Spring MVC 的数据绑定和表单标签,包括数据绑定的原理以及如何使用表单标签。然后给出了一个数据绑定应用示例,大致演示了数据绑定在实际开发中的使用。最后讲解了 JSON 数据交互的使用,使读者了解了 JSON 数据的组织结构。

习　题　5

1. 举例说明数据绑定的优点。
2. Spring MVC 有哪些表单标签?其中可以绑定集合数据的标签有哪些?

拦 截 器

学习目的与要求

本章主要介绍拦截器的概念、原理以及实际应用。通过本章的学习,理解拦截器的原理,掌握拦截器的实际应用。

本章主要内容

- 拦截器的定义。
- 拦截器的配置。
- 拦截器的执行流程。

开发一个网站时,可能有这样的需求:某些页面只希望几个特定的用户浏览。对于这样的访问权限控制,该如何实现呢?拦截器就可以实现上述需求。在 Struts 2 框架中,拦截器是重要的组成部分,Spring MVC 框架也提供了拦截器功能。本章将针对 Spring MVC 中拦截器的使用进行详细讲解。

6.1 拦截器概述

Spring MVC 的拦截器（Interceptor）与 Java Servlet 的过滤器（Filter）类似，它主要用于拦截用户的请求，并做相应的处理。通常应用在权限验证、记录请求信息的日志、判断用户是否登录等功能上。

6.1.1 拦截器的定义

在 Spring MVC 框架中定义一个拦截器，需要对拦截器进行定义和配置。定义一个拦截器可以通过两种方式：一种是通过实现 HandlerInterceptor 接口或继承 HandlerInterceptor 接口的实现类来定义；另一种是通过实现 WebRequestInterceptor 接口或继承 WebRequestInterceptor 接口的实现类来定义。本章以实现 HandlerInterceptor 接口的定义方式为例讲解自定义拦截器的使用方法。示例代码如下：

```
package interceptor;
import javax.servlet.http.HttpServletRequest;
import javax.servlet.http.HttpServletResponse;
import org.springframework.web.servlet.HandlerInterceptor;
import org.springframework.stereotype.Component;
import org.springframework.web.servlet.ModelAndView;
@Component
public class TestInterceptor implements HandlerInterceptor{
    @Override
    public boolean preHandle(HttpServletRequest request, HttpServletResponse response, Object handler)
            throws Exception {
        System.out.println("preHandle 方法在控制器的处理请求方法前执行");
        /**返回 true 表示继续向下执行,返回 false 表示中断后续操作 */
        return true;
    }
    @Override
     public void postHandle (HttpServletRequest request, HttpServletResponse response, Object handler,
            ModelAndView modelAndView) throws Exception {
        System.out.println("postHandle 方法在控制器的处理请求方法调用之后,解析视图之前执行");
    }
    @Override
    public void afterCompletion(HttpServletRequest request, HttpServletResponse response, Object handler, Exception ex)
            throws Exception {
        System.out.println("afterCompletion 方法在控制器的处理请求方法执行完成后
```

执行,即视图渲染结束之后执行");
    }
}

上述拦截器的定义实现了 HandlerInterceptor 接口,并实现了接口中的三个方法。有关这三个方法的描述如下。

preHandle()方法:该方法在控制器的处理请求方法前执行,其返回值表示是否中断后续操作。返回 true 表示继续向下执行,返回 false 表示中断后续操作。

postHandle()方法:该方法在控制器的处理请求方法调用之后、解析视图之前执行。可以通过此方法对请求域中的模型和视图作进一步的修改。

afterCompletion()方法:该方法在控制器的处理请求方法执行完成后执行,即视图渲染结束后执行。可以通过此方法实现一些资源清理、记录日志信息等工作。

6.1.2 拦截器的配置

1. 在 Spring MVC 的 Java 配置类中配置拦截器

让自定义的拦截器生效,可以在 Spring MVC 的 Java 配置类中配置,配置示例代码如下:

```
/**
 * 配置拦截器
 */
@Override
public void addInterceptors(InterceptorRegistry registry) {
    //可配置多个拦截器,testInterceptor 针对 gotoTest 请求有效
    registry.addInterceptor(testInterceptor).addPathPatterns("/gotoTest");
    //registry.addInterceptor(testInterceptor).addPathPatterns("/**");
}
```

2. 在 Spring MVC 的 XML 配置文件中配置拦截器

让自定义的拦截器生效,可以在 Spring MVC 的 XML 配置文件中配置,配置示例代码如下:

```
<!--配置拦截器 -->
<mvc:interceptors>
    <!--配置一个全局拦截器,拦截所有请求 -->
    <bean class="interceptor.TestInterceptor"/>
    <mvc:interceptor>
        <!--配置拦截器作用的路径 -->
        <mvc:mapping path="/**"/>
        <!--配置不需要拦截作用的路径 -->
        <mvc:exclude-mapping path=""/>
        <!--定义在<mvc:interceptor>元素中,表示匹配指定路径的请求才进行拦截 -->
```

```
            <bean class="interceptor.Interceptor1"/>
        </mvc:interceptor>
        <mvc:interceptor>
            <!--配置拦截器作用的路径 -->
            <mvc:mapping path="/gotoTest"/>
            <!--定义在<mvc:interceptor>元素中,表示匹配指定路径的请求才进行拦截 -->
            <bean class="interceptor.Interceptor2"/>
        </mvc:interceptor>
    </mvc:interceptors>
```

在上述示例代码中,<mvc:interceptors>元素用于配置一组拦截器,其子元素<bean>定义的是全局拦截器,即拦截所有的请求。<mvc:interceptor>元素中定义的是指定路径的拦截器,其子元素<mvc:mapping>用于配置拦截器作用的路径,该路径在其属性 path 中定义。如上述示例代码中,path 的属性值"/ * *"表示拦截所有路径,"/gotoTest"表示拦截所有以"/gotoTest"结尾的路径。如果在请求路径中包含不需要拦截的内容,可以通过<mvc:exclude-mapping>子元素配置。

需要注意的是,<mvc:interceptor>元素的子元素必须按照<mvc:mapping .../><mvc:exclude-mapping .../><bean .../>的顺序配置。

6.2 拦截器的执行流程

6.2.1 单个拦截器的执行流程

在配置类中,如果只定义了一个拦截器,程序将首先执行拦截器类中的 preHandle()方法,如果该方法返回 true,程序将继续执行控制器中处理请求的方法,否则中断执行。如果 preHandle()方法返回 true,并且控制器中处理请求的方法执行后返回视图前,将执行 postHandle()方法。返回视图后,才执行 afterCompletion()方法。下面通过一个应用 ch6_1 演示拦截器的执行流程。

【例 6-1】 测试单个拦截器的执行流程。

具体实现步骤如下。

1. 创建 Maven 项目并添加依赖的 JAR 包

创建一个名为 ch6_1 的 Maven 项目,并通过 pom.xml 文件添加项目的相关依赖 spring-webmvc。pom.xml 文件的内容如下:

```
<project xmlns="http://maven.apache.org/POM/4.0.0"
    xmlns:xsi="http://www.w3.org/2001/XMLSchema-instance"
    xsi:schemaLocation="http://maven.apache.org/POM/4.0.0
    https://maven.apache.org/xsd/maven-4.0.0.xsd">
    <modelVersion>4.0.0</modelVersion>
```

```xml
        <groupId>com.cn.chenheng</groupId>
        <artifactId>ch6_1</artifactId>
        <version>0.0.1-SNAPSHOT</version>
        <packaging>war</packaging>
        <properties>
            <!--spring 版本号 -->
            <spring.version>5.2.3.RELEASE</spring.version>
        </properties>
        <dependencies>
            <dependency>
                <groupId>org.springframework</groupId>
                <artifactId>spring-webmvc</artifactId>
                <version>${spring.version}</version>
            </dependency>
        </dependencies>
</project>
```

2. 创建 Web 配置类

在 ch6_1 的 src/main/java 目录下创建一个名为 config 的包,并在该包中创建一个名为 WebConfig 的 Web 配置类。在 Web 配置类中注册 Spring MVC 的 DispatcherServlet,注册字符编码过滤器。Web 配置类的代码如下:

```
package config;
import javax.servlet.ServletContext;
import javax.servlet.ServletException;
import javax.servlet.ServletRegistration.Dynamic;
import org.springframework.web.WebApplicationInitializer;
import org.springframework.web.context.support.
AnnotationConfigWebApplicationContext;
import org.springframework.web.filter.CharacterEncodingFilter;
import org.springframework.web.servlet.DispatcherServlet;
public class WebConfig implements WebApplicationInitializer{
    @Override
    public void onStartup(ServletContext arg0) throws ServletException {
        AnnotationConfigWebApplicationContext ctx
        =new AnnotationConfigWebApplicationContext();
        ctx.register(SpringMVCConfig.class);//注册 Spring MVC 的 Java 配置类
                                            //SpringMVCConfig
        ctx.setServletContext(arg0);//和当前 ServletContext 关联
        /**
         * 注册 Spring MVC 的 DispatcherServlet
         */
        Dynamic servlet =arg0.addServlet("dispatcher", new DispatcherServlet(ctx));
```

```
            servlet.addMapping("/");
            servlet.setLoadOnStartup(1);
            /**
             * 注册字符编码过滤器
             */
            javax.servlet.FilterRegistration.Dynamic filter =
                    arg0.addFilter("characterEncodingFilter",
CharacterEncodingFilter.class);
            filter.setInitParameter("encoding", "UTF-8");
            filter.addMappingForUrlPatterns(null, false, "/*");
    }
}
```

3. 创建控制器类

在 ch6_1 的 src/main/java 目录下创建一个名为 controller 的包,并在该包中创建控制器类 InterceptorController,代码如下:

```
package controller;
import org.springframework.stereotype.Controller;
import org.springframework.web.bind.annotation.RequestMapping;
@Controller
public class InterceptorController {
    @RequestMapping("/gotoTest")
    public String gotoTest() {
        System.out.println("正在测试拦截器,执行控制器的处理请求方法中");
        return "test";
    }
}
```

4. 创建拦截器类

在 ch6_1 的 src/main/java 目录下创建一个名为 interceptor 的包,并在该包中创建拦截器类 TestInterceptor,代码与 6.1.1 小节的示例代码相同。

5. 创建 Spring MVC 的 Java 配置类

在 config 包中创建一个名为 SpringMVCConfig 的 Java 配置类。在 SpringMVCConfig 配置类中配置视图解析器、静态资源及拦截器。该配置类的代码如下:

```
package config;
import org.springframework.beans.factory.annotation.Autowired;
import org.springframework.context.annotation.Bean;
import org.springframework.context.annotation.ComponentScan;
import org.springframework.context.annotation.Configuration;
```

```java
import org.springframework.web.servlet.HandlerInterceptor;
import org.springframework.web.servlet.config.annotation.EnableWebMvc;
import org.springframework.web.servlet.config.annotation.InterceptorRegistry;
import org.springframework.web.servlet.config.annotation.ResourceHandlerRegistry;
import org.springframework.web.servlet.config.annotation.WebMvcConfigurer;
import org.springframework.web.servlet.view.InternalResourceViewResolver;
@Configuration
@EnableWebMvc // 开启 Spring MVC 的支持
@ComponentScan(basePackages = { "controller", "interceptor"}) // 扫描基本包
public class SpringMVCConfig implements WebMvcConfigurer {
    @Autowired
    private HandlerInterceptor testInterceptor;//依赖注入自定义拦截器
    /**
     * 配置视图解析器
     */
    @Bean
    public InternalResourceViewResolver getViewResolver() {
        InternalResourceViewResolver viewResolver = new InternalResourceViewResolver();
        viewResolver.setPrefix("/WEB-INF/jsp/");
        viewResolver.setSuffix(".jsp");
        return viewResolver;
    }
    /**
     * 配置静态资源(不需要 DispatcherServlet 转发的请求)
     */
    @Override
    public void addResourceHandlers(ResourceHandlerRegistry registry) {
        registry.addResourceHandler("/static/**").addResourceLocations
        ("/static/");
    }
    /**
     * 配置拦截器
     */
    @Override
    public void addInterceptors(InterceptorRegistry registry) {
        //可配置多个拦截器,testInterceptor 针对 gotoTest 请求有效
        registry.addInterceptor(testInterceptor).addPathPatterns("/gotoTest");
        //registry.addInterceptor(testInterceptor).addPathPatterns("/**");
    }
}
```

6. 创建视图 JSP 文件

在 src/main/webapp/WEB-INF 目录下创建一个 jsp 文件夹,并在该文件夹中创建

一个 JSP 文件 test.jsp，代码如下：

```
<%@page language="java" contentType="text/html; charset=UTF-8" pageEncoding="UTF-8"%>
<%
String path =request.getContextPath();
String basePath = request.getScheme()+"://"+request.getServerName()+":"+request.getServerPort()+path+"/";
%>
<!DOCTYPE html>
<html>
<head>
<base href="<%=basePath%>">
<meta charset="UTF-8">
<title>Insert title here</title>
</head>
<body>
    视图
    <%System.out.println("视图渲染结束。");%>
</body>
</html>
```

7. 测试拦截器

首先，将应用 ch6_1 发布到 Tomcat 服务器，并启动 Tomcat 服务器。然后，通过地址 http://localhost:8080/ch6_1/gotoTest 测试拦截器。程序正确执行后，控制台的输出结果如图 6.1 所示。

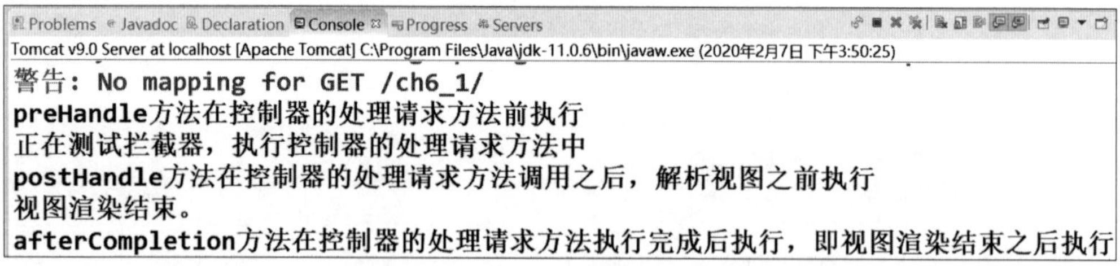

```
警告: No mapping for GET /ch6_1/
preHandle方法在控制器的处理请求方法前执行
正在测试拦截器，执行控制器的处理请求方法中
postHandle方法在控制器的处理请求方法调用之后，解析视图之前执行
视图渲染结束。
afterCompletion方法在控制器的处理请求方法执行完成后执行，即视图渲染结束之后执行
```

图 6.1　单个拦截器的执行过程

6.2.2　多个拦截器的执行流程

在 Web 应用中，通常有多个拦截器同时工作，这时它们的 preHandle()方法将按照配置文件或配置类中拦截器的配置顺序执行，而它们的 postHandle()方法和 afterCompletion()方法则按照配置顺序的反序执行。

下面通过修改 6.2.1 小节的应用 ch6_1 来演示多个拦截器的执行流程。

【例 6-2】　演示多个拦截器的执行流程。

具体实现步骤如下。

1. 创建多个拦截器

在 ch6_1 应用的 interceptor 包中创建两个拦截器类 Interceptor1 和 Interceptor2。
Interceptor1 类的代码如下:

```
package interceptor;
import javax.servlet.http.HttpServletRequest;
import javax.servlet.http.HttpServletResponse;
import org.springframework.web.servlet.HandlerInterceptor;
import org.springframework.stereotype.Component;
import org.springframework.web.servlet.ModelAndView;
@Component
public class Interceptor1 implements HandlerInterceptor{
    @Override
    public boolean preHandle(HttpServletRequest request, HttpServletResponse response, Object handler)
            throws Exception {
        System.out.println("Interceptor1 preHandle 方法执行");
        /**返回 true 表示继续向下执行,返回 false 表示中断后续的操作*/
        return true;
    }
    @Override
     public void postHandle (HttpServletRequest request, HttpServletResponse response, Object handler,
            ModelAndView modelAndView) throws Exception {
        System.out.println("Interceptor1 postHandle 方法执行");
    }
    @Override
    public void afterCompletion(HttpServletRequest request, HttpServletResponse response, Object handler, Exception ex)
            throws Exception {
        System.out.println("Interceptor1 afterCompletion 方法执行");
    }
}
```

Interceptor2 类的代码如下:

```
package interceptor;
import javax.servlet.http.HttpServletRequest;
import javax.servlet.http.HttpServletResponse;
import org.springframework.web.servlet.HandlerInterceptor;
import org.springframework.stereotype.Component;
import org.springframework.web.servlet.ModelAndView;
@Component
```

```java
public class Interceptor2 implements HandlerInterceptor{
    @Override
    public boolean preHandle(HttpServletRequest request, HttpServletResponse response, Object handler)
            throws Exception {
        System.out.println("Interceptor2 preHandle 方法执行");
        /**返回 true 表示继续向下执行,返回 false 表示中断后续的操作*/
        return true;
    }
    @Override
    public void postHandle(HttpServletRequest request, HttpServletResponse response, Object handler,
            ModelAndView modelAndView) throws Exception {
        System.out.println("Interceptor2 postHandle 方法执行");
    }
    @Override
    public void afterCompletion(HttpServletRequest request, HttpServletResponse response, Object handler, Exception ex)
            throws Exception {
        System.out.println("Interceptor2 afterCompletion 方法执行");
    }
}
```

2. 配置拦截器

在 Spring MVC 的配置类 SpringMVCConfig 中配置拦截器 Interceptor1 和 Interceptor2。配置代码如下：

```java
package config;
import org.springframework.beans.factory.annotation.Autowired;
import org.springframework.context.annotation.Bean;
import org.springframework.context.annotation.ComponentScan;
import org.springframework.context.annotation.Configuration;
import org.springframework.web.servlet.HandlerInterceptor;
import org.springframework.web.servlet.config.annotation.EnableWebMvc;
import org.springframework.web.servlet.config.annotation.InterceptorRegistry;
import org.springframework.web.servlet.config.annotation.ResourceHandlerRegistry;
import org.springframework.web.servlet.config.annotation.WebMvcConfigurer;
import org.springframework.web.servlet.view.InternalResourceViewResolver;
@Configuration
@EnableWebMvc // 开启 Spring MVC 的支持
@ComponentScan(basePackages ={ "controller", "interceptor"}) // 扫描基本包
public class SpringMVCConfig implements WebMvcConfigurer {
    @Autowired
```

```java
    private HandlerInterceptor testInterceptor;//依赖注入自定义拦截器
    @Autowired
    private HandlerInterceptor interceptor1;//依赖注入自定义拦截器
    @Autowired
    private HandlerInterceptor interceptor2;//依赖注入自定义拦截器
    /**
     * 配置视图解析器
     */
    @Bean
    public InternalResourceViewResolver getViewResolver() {
        InternalResourceViewResolver viewResolver =new InternalResourceViewResolver();
        viewResolver.setPrefix("/WEB-INF/jsp/");
        viewResolver.setSuffix(".jsp");
        return viewResolver;
    }
    /**
     * 配置静态资源(不需要 DispatcherServlet 转发的请求)
     */
    @Override
    public void addResourceHandlers(ResourceHandlerRegistry registry) {
        registry.addResourceHandler("/static/**").addResourceLocations("/static/");
    }
    /**
     * 配置拦截器
     */
    @Override
    public void addInterceptors(InterceptorRegistry registry) {
        //配置全局拦截器
        registry.addInterceptor(testInterceptor);
        //表示匹配指定路径的请求才进行拦截
        registry.addInterceptor(interceptor1).addPathPatterns("/**");
        //表示匹配指定路径的请求才进行拦截
        registry.addInterceptor(interceptor2).addPathPatterns("/gotoTest");
    }
}
```

3. 测试多个拦截器

首先,将应用 ch6_1 发布到 Tomcat 服务器,并启动 Tomcat 服务器。然后,通过地址 http://localhost:8080/ch6_1/gotoTest 测试拦截器。程序正确执行后,控制台的输出结果如图 6.2 所示。

```
2月 07, 2020 4:10:04 下午 org.springframework.web.servlet.DispatcherServlet
警告: No mapping for GET /ch6_1/
preHandle方法在控制器的处理请求方法前执行
Interceptor1 preHandle方法执行
Interceptor2 preHandle方法执行
正在测试拦截器，执行控制器的处理请求方法中
Interceptor2 postHandle方法执行
Interceptor1 postHandle方法执行
postHandle方法在控制器的处理请求方法调用之后，解析视图之前执行
视图渲染结束。
Interceptor2 afterCompletion方法执行
Interceptor1 afterCompletion方法执行
afterCompletion方法在控制器的处理请求方法执行完成后执行，即视图渲染结束之后执行
```

图 6.2　多个拦截器的执行过程

6.3　应用案例——用户登录权限验证

本节将通过拦截器来完成一个用户登录权限验证的 Spring MVC Web 应用 ch6_2。具体要求为：只有成功登录的用户才能访问系统的主页面 main.jsp，如果没有成功登录而直接访问主页面，则拦截器将拦截请求，并转到登录页面 login.jsp。当成功登录的用户在系统主页面中单击"退出"链接时，回到登录页面。

【例 6-3】　用户登录权限验证。

具体实现步骤如下。

1. 创建 Maven 项目并添加依赖的 JAR 包

创建一个名为 ch6_2 的 Maven 项目，并通过 pom.xml 文件添加项目的相关依赖 spring-webmvc。pom.xml 文件的内容与【例 6-1】的相同，不再赘述。

2. 创建 POJO 类

在 ch6_2 的 src/main/java 目录中创建一个名为 pojo 的包，并在该包中创建 User 类。具体代码如下：

```
package pojo;
public class User {
    private String uname;
    private String upwd;
    //省略 setter 和 getter 方法
}
```

3. 创建控制器类

在 ch6_2 的 src/main/java 目录中创建一个名为 controller 的包，并在该包中创建控

制器类 UserController。具体代码如下：

```java
package controller;
import javax.servlet.http.HttpSession;
import org.springframework.stereotype.Controller;
import org.springframework.ui.Model;
import org.springframework.web.bind.annotation.RequestMapping;
import pojo.User;
@Controller
public class UserController {
    /**
     * 登录页面初始化
     */
    @RequestMapping("/toLogin")
    public String initLogin() {
        return "login";
    }
    /**
     * 处理登录功能
     */
    @RequestMapping("/login")
    public String login(User user, Model model,HttpSession session) {
        System.out.println(user.getUname());
        if("chenheng".equals(user.getUname()) &&
                "123456".equals(user.getUpwd())) {
            //登录成功,将用户信息保存到session对象中
            session.setAttribute("user", user);
            //重定向到主页面的跳转方法
            return "redirect:main";
        }
        model.addAttribute("msg","用户名或密码错误,请重新登录!");
        return "login";
    }
    /**
     * 跳转到主页面
     */
    @RequestMapping("/main")
    public String toMain() {
        return "main";
    }
    /**
     * 退出登录
     */
    @RequestMapping("/logout")
```

```java
public String logout(HttpSession session) {
    //清除 Session
    session.invalidate();
    return "login";
}
```

4. 创建拦截器类

在 ch6_2 的 src/main/java 目录中创建一个名为 interceptor 的包,并在该包中创建拦截器类 LoginInterceptor。具体代码如下:

```java
package interceptor;
import javax.servlet.http.HttpServletRequest;
import javax.servlet.http.HttpServletResponse;
import javax.servlet.http.HttpSession;
import org.springframework.stereotype.Component;
import org.springframework.web.servlet.HandlerInterceptor;
@Component
public class LoginInterceptor implements HandlerInterceptor{
    @Override
    public boolean preHandle(HttpServletRequest request, HttpServletResponse
    response, Object handler) throws Exception {
        //获取请求的 URL
        String url = request.getRequestURI();
        //login.jsp 或登录请求放行,不拦截
        if(url.indexOf("/toLogin") >= 0 || url.indexOf("/login") >= 0) {
            return true;
        }
        //获取 Session
        HttpSession session = request.getSession();
        Object obj = session.getAttribute("user");
        if(obj != null)
            return true;
        //没有登录且不是登录页面,转到登录页面,并给出提示错误信息
        request.setAttribute("msg", "还没登录,请先登录!");
        request.getRequestDispatcher("/WEB-INF/jsp/login.jsp").forward
        (request, response);
        return false;
    }
}
```

5. 创建 Web 配置类

在 ch6_2 的 src/main/java 目录下创建一个名为 config 的包,并在该包中创建一个名

为 WebConfig 的 Web 配置类。在 Web 配置类中注册 Spring MVC 的 DispatcherServlet,注册字符编码过滤器。Web 配置类的代码与【例 6-1】的相同,不再赘述。

6. 创建 Spring MVC 的 Java 配置类

在 config 包中创建一个名为 SpringMVCConfig 的 Java 配置类。在 SpringMVCConfig 配置类中配置视图解析器、静态资源以及拦截器。该配置类的代码如下:

```
package config;
import org.springframework.beans.factory.annotation.Autowired;
import org.springframework.context.annotation.Bean;
import org.springframework.context.annotation.ComponentScan;
import org.springframework.context.annotation.Configuration;
import org.springframework.web.servlet.HandlerInterceptor;
import org.springframework.web.servlet.config.annotation.EnableWebMvc;
import org.springframework.web.servlet.config.annotation.InterceptorRegistry;
import org.springframework.web.servlet.config.annotation.ResourceHandlerRegistry;
import org.springframework.web.servlet.config.annotation.WebMvcConfigurer;
import org.springframework.web.servlet.view.InternalResourceViewResolver;
@Configuration
@EnableWebMvc // 开启 Spring MVC 的支持
@ComponentScan(basePackages ={ "controller", "interceptor"}) // 扫描基本包
public class SpringMVCConfig implements WebMvcConfigurer {
    @Autowired
    private HandlerInterceptor loginInterceptor;//依赖注入自定义拦截器
    /**
     * 配置视图解析器
     */
    @Bean
    public InternalResourceViewResolver getViewResolver() {
        InternalResourceViewResolver viewResolver =new
                InternalResourceViewResolver();
        viewResolver.setPrefix("/WEB-INF/jsp/");
        viewResolver.setSuffix(".jsp");
        return viewResolver;
    }
    /**
     * 配置静态资源(不需要 DispatcherServlet 转发的请求)
     */
    @Override
    public void addResourceHandlers(ResourceHandlerRegistry registry) {
        registry.addResourceHandler("/static/**").addResourceLocations("/static/");
    }
    /**
     * 配置拦截器
```

```
    */
    @Override
    public void addInterceptors(InterceptorRegistry registry) {
        //addPathPatterns 表示匹配指定路径的请求才进行拦截;excludePathPatterns 表
          示不拦截的路径。
        registry.addInterceptor(loginInterceptor).addPathPatterns("/**").
        excludePathPatterns("/static/**");
    }
}
```

7. 创建视图 JSP 页面

在 ch6_2 的 src/main/webapp/WEB-INF 目录下创建文件夹 jsp,并在该文件夹中创建 login.jsp 和 main.jsp。

login.jsp 的代码如下:

```
<%@page language="java" contentType="text/html; charset=UTF-8" pageEncoding="UTF-8"%>
<%
String path = request.getContextPath();
String basePath = request.getScheme()+"://"+request.getServerName()+":"+request.getServerPort()+path+"/";
%>
<!DOCTYPE html>
<html>
<head>
<base href="<%=basePath%>">
<meta charset="UTF-8">
<title>Insert title here</title>
<link rel="stylesheet" href="static/css/bootstrap.min.css" />
</head>
<body>
    <div class="container">
        <div>
                <h4>${msg}</h4>
        </div>
        <div class="row">
            <div class="col-md-6 col-sm-6">
                <form class="form-horizontal" action="login" method="post">
                    <div class="form-group">
                        <div class="input-group col-md-6">
                            <span class="input-group-addon">
                                <i class="glyphicon glyphicon-pencil"></i>
                            </span>
```

```html
                    <input class="form-control" type="text"
                    name="uname" placeholder="请输入用户名"/>
                </div>
            </div>
            <div class="form-group">
                <div class="input-group col-md-6">
                    <span class="input-group-addon">
                        <i class="glyphicon glyphicon-pencil"></i>
                    </span>
                    <input class="form-control" type="password"
                     name="upwd" placeholder="请输入密码"/>
                </div>
            </div>
            <div class="form-group">
                <div class="col-md-6">
                    <div class="btn-group btn-group-justified">
                        <div class="btn-group">
                            <button type="submit" class="btn btn-success">
                                <span class="glyphicon glyphicon-share">
                                </span>
                                 登录
                            </button>
                        </div>
                    </div>
                </div>
            </div>
        </form>
    </div>
  </div>
</body>
</html>
```

main.jsp 的代码如下：

```jsp
<%@page language="java" contentType="text/html; charset=UTF-8" pageEncoding="UTF-8"%>
<%
String path =request.getContextPath();
String basePath = request.getScheme()+"://"+request.getServerName()+":"+
request.getServerPort()+path+"/";
%>
<!DOCTYPE html>
<html>
<head>
```

```
<base href="<%=basePath%>">
<meta charset="UTF-8">
<title>Insert title here</title>
</head>
<body>
    当前用户:${user.uname }<br>
    <a href="logout">退出</a>
</body>
</html>
```

8. 发布并测试应用

首先,将应用 ch6_2 发布到 Tomcat 服务器,并启动 Tomcat 服务器。然后,通过地址 http://localhost:8080/ch6_2/main 测试应用,运行效果如图 6.3 所示。

图 6.3 没有登录直接访问主页面的效果

从图 6.3 可以看出,当用户没有登录而直接访问系统主页面时,请求将被登录拦截器拦截,返回登录页面,并提示信息。如果用户在用户名输入框中输入 chenheng,密码框中输入 123456,单击"登录"按钮后,浏览器的显示结果如图 6.4 所示。如果输入的用户名或密码错误,浏览器的显示结果如图 6.5 所示。

图 6.4 成功登录的效果

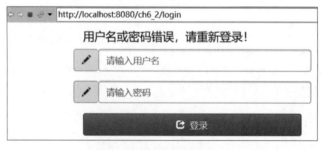

图 6.5 用户名或密码错误

当单击图 6.4 中的"退出"链接后,系统将从主页面返回到登录页面。

6.4 本章小结

本章首先讲解了在 Spring MVC 应用中如何定义和配置拦截器;然后详细讲解了拦截器的执行流程,包括单个和多个拦截器的执行流程;最后通过用户登录权限验证的应用案例演示了拦截器的实际应用。

习 题 6

1. 在 Spring MVC 框架中,如何自定义拦截器?又如何配置自定义拦截器?
2. 请简述单个拦截器和多个拦截器的执行流程。

数据验证

学习目的与要求

本章重点讲解 Spring MVC 框架的输入验证体系。通过本章的学习,理解输入验证的流程,能够利用 Spring 的自带验证框架和 JSR 303(Java 验证规范)对数据进行验证。

本章主要内容

- 数据验证概述。
- Spring 验证。
- JSR 303 验证。

用户的输入一般是随意的,为了保证数据的合法性,数据验证是所有 Web 应用必须处理的问题。在 Spring MVC 框架中,有两种方法可以验证输入数据:一是利用 Spring 自带的验证框架;二是利用 JSR 303 实现。

7.1 数据验证概述

数据验证分为客户端验证和服务器端验证,客户端验证主要是过滤正常用户的误操作,一般通过 JavaScript 代码实现;服务器端验证是整个应用阻止非法数据的最后防线,主要在应用中编程实现。

7.1.1 客户端验证

在大多数情况下,使用 JavaScript 进行客户端验证的步骤如下:
- 编写验证函数。
- 在提交表单的事件中调用验证函数。
- 根据验证函数判断是否进行表单提交。

客户端验证可以过滤用户的误操作,是第一道防线,一般使用 JavaScript 代码实现。仅有客户端验证还是不够的。攻击者还可以绕过客户端验证直接进行非法输入,这样可能会引起系统的异常。所以必须加上服务器端的验证。

7.1.2 服务器端验证

Spring MVC 的 Converter 和 Formatter 在进行类型转换时,是将输入数据转换成领域对象的属性值(一种 Java 类型),一旦成功,服务器端验证器就会介入。也就是说,在 Spring MVC 框架中,先进行数据类型转换,再进行服务器端验证。

服务器端验证对于系统的安全性、完整性、健壮性起到了至关重要的作用。在 Spring MVC 框架中,可以利用 Spring 自带的验证框架验证数据,也可以利用 JSR 303 实现数据验证。

7.2 Spring 验证器

7.2.1 Validator 接口

创建自定义 Spring 验证器,需要实现 org.springframework.validation.Validator 接口。该接口有以下两个接口方法:

```
boolean supports(Class<?>klass)
void validate(Object object, Errors errors)
```

supports 方法返回为 true 时,验证器可以处理指定的 Class。validate 方法的功能是验证目标对象 object,并将验证错误消息存入 Errors 对象。

往 Errors 对象中存入错误消息的方法是 reject 或 rejectValue 方法。这两个方法的部分重载方法如下:

```
void reject(String errorCode)
```

```
void reject(String errorCode, String defaultMessage)
void rejectValue(String field, String errorCode)
void rejectValue(String field, String errorCode, String defaultMessage)
```

在一般情况下，只需给 reject 或 rejectValue 方法一个错误代码，Spring MVC 框架就会在消息属性文件中查找错误代码，获取相应错误消息。具体示例如下：

```
if(goods.getGprice() >100 || goods.getGprice() <0){
    errors.rejectValue("gprice", "gprice.invalid");//gprice.invalid 为错误
                                                    //代码
}
```

7.2.2 ValidationUtils 类

org.springframework.validation.ValidationUtils 是一个工具类，该类中有几个方法可以帮助判定值是否为空。

例如：

```
if(goods.getGname() ==null || goods.getGname().isEmpty()){
    errors.rejectValue("gname", "goods.gname.required")
}
```

上述 if 语句可以使用 ValidationUtils 类的 rejectIfEmpty 方法，代码如下：

```
// errors 为 Errors 对象
// gname 为 goods 对象的属性
ValidationUtils.rejectIfEmpty(errors, "gname", "goods.gname.required");
```

再如：

```
if(goods.getGname() ==null || goods.getGname().trim().isEmpty()){
    errors.rejectValue("gname", "goods.gname.required")
}
```

上述 if 语句可以编写成以下语句：

```
//gname 为 goods 对象的属性
ValidationUtils.rejectIfEmptyOrWhitespace(errors, "gname", "goods.gname.required");
```

7.2.3 验证示例

本节使用一个应用 ch7_1 讲解 Spring 验证器的编写及使用。该应用中有 1 个数据输入页面 addGoods.jsp，效果如图 7.1 所示；有 1 个数据显示页面 goodsList.jsp，效果如图 7.2 所示。

编写一个实现 org.springframework.validation.Validator 接口的验证器类 GoodsValidator，验证要求如下：

（1）商品名称和商品详情不能为空。
（2）商品价格在 0～100 之间。
（3）创建日期不能在系统日期之后。

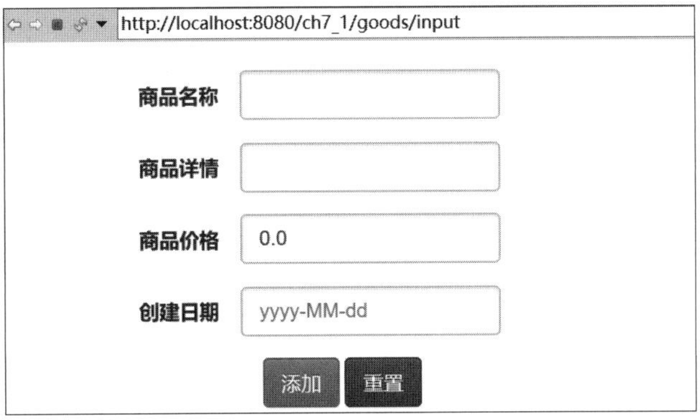

图 7.1　数据输入页面

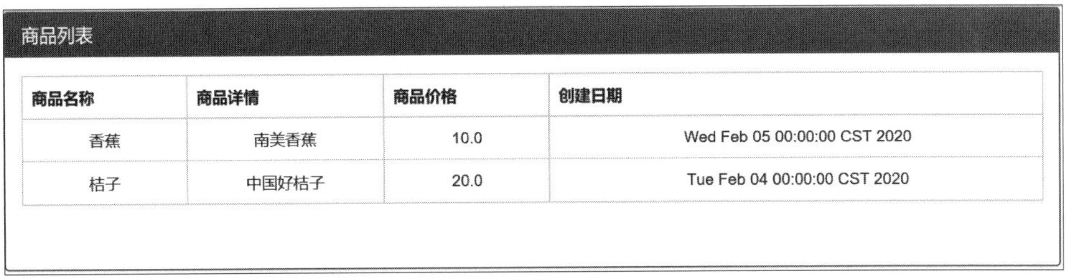

商品列表

商品名称	商品详情	商品价格	创建日期
香蕉	南美香蕉	10.0	Wed Feb 05 00:00:00 CST 2020
桔子	中国好桔子	20.0	Tue Feb 04 00:00:00 CST 2020

图 7.2　数据显示页面

【例 7-1】 Spring 验证器的编写及使用。
具体实现步骤如下。

1. 创建 Maven 项目并添加相关依赖

创建一个名为 ch7_1 的 Maven 项目，并通过 pom.xml 文件添加项目的相关依赖 spring-webmvc 和 jstl。pom.xml 文件的内容如下：

```
<project xmlns="http://maven.apache.org/POM/4.0.0"
    xmlns:xsi="http://www.w3.org/2001/XMLSchema-instance"
    xsi:schemaLocation="http://maven.apache.org/POM/4.0.0
https://maven.apache.org/xsd/maven-4.0.0.xsd">
    <modelVersion>4.0.0</modelVersion>
    <groupId>com.cn.chenheng</groupId>
    <artifactId>ch7_1</artifactId>
    <version>0.0.1-SNAPSHOT</version>
    <packaging>war</packaging>
    <properties>
        <!-- spring 版本号 -->
```

```xml
        <spring.version>5.2.3.RELEASE</spring.version>
    </properties>
    <dependencies>
        <dependency>
            <groupId>org.springframework</groupId>
            <artifactId>spring-webmvc</artifactId>
            <version>${spring.version}</version>
        </dependency>
        <dependency>
            <groupId>javax.servlet</groupId>
            <artifactId>jstl</artifactId>
            <version>1.2</version>
        </dependency>
    </dependencies>
</project>
```

2. 编写数据输入页面

在应用 ch7_1 的 src/main/webapp/WEB-INF 目录下创建文件夹 jsp,并在该文件夹中创建数据输入页面 addGoods.jsp。addGoods.jsp 的代码如下:

```jsp
<%@ page language="java" contentType="text/html; charset=UTF-8" pageEncoding="UTF-8"%>
<%@ taglib prefix="form" uri="http://www.springframework.org/tags/form" %>
<%
String path = request.getContextPath();
String basePath = request.getScheme()+"://"+request.getServerName()+":"+request.getServerPort()+path+"/";
%>
<!DOCTYPE html>
<html>
<head>
<base href="<%=basePath%>">
<meta charset="UTF-8">
<title>Insert title here</title>
<link rel="stylesheet" href="static/css/bootstrap.min.css" />
</head>
<body>
<form:form cssClass="form-horizontal" action="goods/save" method="post" modelAttribute="goods" >
    <br>
    <div class="form-group">
        <label class="col-sm-2 col-md-2 control-label">商品名称</label>
        <div class="input-group col-md-2">
```

```
                <form:input cssClass="form-control" path="gname"/>
            </div>
        </div>
        <div class="form-group">
            <label class="col-sm-2 col-md-2 control-label">商品详情</label>
            <div class="input-group col-md-2">
                <form:input cssClass="form-control" path="gdescription"/>
            </div>
        </div>
        <div class="form-group">
            <label class="col-sm-2 col-md-2 control-label">商品价格</label>
            <div class="input-group col-md-2">
                <form:input cssClass="form-control" path="gprice"/>
            </div>
        </div>
        <div class="form-group">
            <label class="col-sm-2 col-md-2 control-label">创建日期</label>
            <div class="input-group col-md-2">
                <form:input cssClass="form-control" placeholder="yyyy-MM-dd"
                    path="gdate"/>
            </div>
        </div>
        <div class="form-group">
            <div class="col-sm-offset-2 col-sm-10">
                <button type="submit"class="btn btn-success" >添加</button>
                <button type="reset" class="btn btn-primary" >重置</button>
            </div>
        </div>
        <div class="form-group">
            <label class="col-sm-3 col-md-4 control-label">
                <!--取出所有验证错误 -->
                <form:errors  path=" * "/>
            </label>
        </div>
    </form:form>
</body>
</html>
```

3. 编写模型类

在应用 ch7_1 的 src/main/java 目录下创建一个名为 pojo 的包，并在该包中定义领域模型类 Goods，封装输入参数。在该类中使用＠DateTimeFormat(pattern＝"yyyy-MM-dd")注解格式化创建日期。模型类 Goods 的具体代码如下：

```
package pojo;
import java.util.Date;
import org.springframework.format.annotation.DateTimeFormat;
public class Goods {
    private String gname;
    private String gdescription;
    private double gprice;
    //日期格式化
    @DateTimeFormat(pattern="yyyy-MM-dd")
    private Date gdate;
    //省略 setter 和 getter 方法
}
```

4. 编写验证器类

在应用 ch7_1 的 src/main/java 目录下创建一个名为 validator 的包,并在该包中编写实现 org.springframework.validation.Validator 接口的验证器类 GoodsValidator,使用 @Component 注解将 GoodsValidator 类声明为验证组件。具体代码如下:

```
package validator;
import java.util.Date;
import org.springframework.stereotype.Component;
import org.springframework.validation.Errors;
import org.springframework.validation.ValidationUtils;
import org.springframework.validation.Validator;
import pojo.Goods;
@Component
public class GoodsValidator implements Validator{
    @Override
    public boolean supports(Class<?>klass) {
        // 要验证的 Model,返回值为 false 则不验证
        return Goods.class.isAssignableFrom(klass);
    }
    @Override
    public void validate(Object object, Errors errors) {
        Goods goods = (Goods)object;//object 要验证的对象
        //goods.gname.required 是错误消息属性文件中的 key
        ValidationUtils.rejectIfEmpty(errors, "gname", "goods.gname.required");
        ValidationUtils.rejectIfEmpty(errors, "gdescription", "goods.gdescription.
                                  required");
        if(goods.getGprice() >100 || goods.getGprice() <0){
            errors.rejectValue("gprice", "gprice.invalid");
        }
        Date goodsDate =goods.getGdate();
```

```java
        //在系统时间之后
        if(goodsDate !=null && goodsDate.after(new Date())){
            errors.rejectValue("gdate", "gdate.invalid");
        }
    }
}
```

5. 编写错误消息属性文件

在应用 ch7_1 的 src/main/resources 目录下创建一个名为 errorMessages.properties 的属性文件。文件内容如下：

```
goods.gname.required=请输入商品名称
goods.gdescription.required=请输入商品详情
gprice.invalid=价格在 0~100
gdate.invalid=创建日期不能在系统日期之后
```

Unicode 编码（Eclipse 带有将汉字转换成 Unicode 编码的功能）的属性文件内容如下：

```
goods.gname.required=\u8BF7\u8F93\u5165\u5546\u54C1\u540D\u79F0\u3002
goods.gdescription.required=\u8BF7\u8F93\u5165\u5546\u54C1\u8BE6\u60C5\u3002
gprice.invalid=\u4EF7\u683C\u57280-100\u4E4B\u95F4\u3002
gdate.invalid=\u521B\u5EFA\u65E5\u671F\u4E0D\u80FD\u5728\u7CFB\u7EDF\u65E5\u671F\u4E4B\u540E\u3002
```

创建完属性文件后，需要告诉 Spring MVC 的 Java 配置类从该文件中获取错误消息。示例代码如下：

```java
/**
 * 配置消息属性文件 errorMessages.properties
 * Bean 的名字必须是 messageSource，这是 Spring 规定的
 */
@Bean("messageSource")
public ResourceBundleMessageSource getMessageSource() {
    ResourceBundleMessageSource rbms =new ResourceBundleMessageSource();
    //默认找 src/main/resources 目录下的 errorMessages.properties
    rbms.setBasename("errorMessages");
    rbms.setDefaultEncoding("UTF-8");
    return rbms;
}
```

6. 编写 Service 层

在应用 ch7_1 的 src/main/java 目录下创建一个名为 service 的包，并在该包中编写一个 GoodsService 接口和 GoodsServiceImpl 实现类。

GoodsService 接口代码如下：

```
package service;
import java.util.ArrayList;
import pojo.Goods;
public interface GoodsService {
    boolean save(Goods g);
    ArrayList<Goods> getGoods();
}
```

GoodsServiceImpl 实现类代码如下：

```
package service;
import java.util.ArrayList;
import org.springframework.stereotype.Service;
import pojo.Goods;
@Service
public class GoodsServiceImpl implements GoodsService{
    //使用静态集合变量 goods 模拟数据库
    private static ArrayList<Goods> goods =new ArrayList<Goods>();
    @Override
    public boolean save(Goods g) {
        goods.add(g);
        return true;
    }
    @Override
    public ArrayList<Goods> getGoods() {
        return goods;
    }
}
```

7. 编写控制器类

在应用 ch7_1 的 src/main/java 目录下创建一个名为 controller 的包，并在该包中编写控制器类 GoodsController，在该类中使用@Autowired 注解注入自定义验证器。另外，控制器类中包含两个处理请求的方法，具体代码如下：

```
package controller;
import org.apache.commons.logging.Log;
import org.apache.commons.logging.LogFactory;
import org.springframework.beans.factory.annotation.Autowired;
import org.springframework.stereotype.Controller;
import org.springframework.ui.Model;
import org.springframework.validation.BindingResult;
import org.springframework.validation.Validator;
import org.springframework.web.bind.annotation.ModelAttribute;
```

```java
import org.springframework.web.bind.annotation.RequestMapping;
import pojo.Goods;
import service.GoodsService;
@Controller
@RequestMapping("/goods")
public class GoodsController {
    // 得到一个用来记录日志的对象,这样打印信息的时候能够标记打印的是哪个类的信息
    private static final Log logger =LogFactory.getLog(GoodsController.class);
    @Autowired
    private GoodsService goodsService;
    @Autowired
    private Validator goodsValidator;
    @RequestMapping("/input")
    public String input(Model model){
        // 如果 model 中没有 goods 属性,addGoods.jsp 会抛出异常,
        // 因为表单标签无法找到 modelAttribute 属性指定的 form
        // backing object
        model.addAttribute("goods", new Goods());
        return "addGoods";
    }
    @RequestMapping("/save")
    public String save(@ModelAttribute Goods goods, Model model, BindingResult
                    result){
        this.goodsValidator.validate(goods, result);//添加验证
        if (result.hasErrors()) {
            return "addGoods";
        }
        goodsService.save(goods);
        logger.info("添加成功");
        model.addAttribute("goodsList", goodsService.getGoods());
        return "goodsList";
    }
}
```

8. 创建 Spring MVC 和 Web 配置类

在应用 ch7_1 的 src/main/java 目录下创建一个名为 config 的包,并在该包中创建 Spring MVC 的 Java 配置类 SpringMVCConfig 和 Web 的 Java 配置类 WebConfig。

在 SpringMVCConfig 类中配置视图解析器、静态资源及消息属性文件。需要注意的是,配置消息属性文件的 Bean 的名字必须是 messageSource,即@Bean("messageSource"),这是 Spring 规定的。具体代码如下:

```java
package config;
import org.springframework.context.annotation.Bean;
```

```java
import org.springframework.context.annotation.ComponentScan;
import org.springframework.context.annotation.Configuration;
import org.springframework.context.support.ResourceBundleMessageSource;
import org.springframework.web.servlet.config.annotation.EnableWebMvc;
import org.springframework.web.servlet.config.annotation.ResourceHandlerRegistry;
import org.springframework.web.servlet.config.annotation.WebMvcConfigurer;
import org.springframework.web.servlet.view.InternalResourceViewResolver;
@Configuration
@EnableWebMvc  // 开启 Spring MVC 的支持
@ComponentScan(basePackages ={ "controller", "service", "validator"})
                                                                    // 扫描基本包
public class SpringMVCConfig implements WebMvcConfigurer {
    /**
     * 配置视图解析器
     */
    @Bean
    public InternalResourceViewResolver getViewResolver() {
        InternalResourceViewResolver viewResolver =new InternalResourceViewResolver();
        viewResolver.setPrefix("/WEB-INF/jsp/");
        viewResolver.setSuffix(".jsp");
        return viewResolver;
    }
    /**
     * 配置消息属性文件 errorMessages.properties
     * Bean 的名字必须是 messageSource,这是 Spring 规定的
     */
    @Bean("messageSource")
    public ResourceBundleMessageSource getMessageSource() {
        ResourceBundleMessageSource rbms =new ResourceBundleMessageSource();
        //默认找 src/main/resources 目录下的 errorMessages.properties
        rbms.setBasename("errorMessages");
        rbms.setDefaultEncoding("UTF-8");
        return rbms;
    }
    /**
     * 配置静态资源(不需要 DispatcherServlet 转发的请求)
     */
    @Override
    public void addResourceHandlers(ResourceHandlerRegistry registry) {
        registry.addResourceHandler("/static/**").addResourceLocations
        ("/static/");
    }
}
```

在 WebConfig 类中注册 Spring MVC 的 DispatcherServlet 和字符编码过滤器。具

体代码如下:

```java
package config;
import javax.servlet.ServletContext;
import javax.servlet.ServletException;
import javax.servlet.ServletRegistration.Dynamic;
import org.springframework.web.WebApplicationInitializer;
import org.springframework.web.context.support.AnnotationConfigWebApplicationContext;
import org.springframework.web.filter.CharacterEncodingFilter;
import org.springframework.web.servlet.DispatcherServlet;
public class WebConfig implements WebApplicationInitializer{
    @Override
    public void onStartup(ServletContext arg0) throws ServletException {
        AnnotationConfigWebApplicationContext ctx
        =new AnnotationConfigWebApplicationContext();
        ctx.register(SpringMVCConfig.class);//注册 Spring MVC 的 Java 配置类
                                            //SpringMVCConfig
        ctx.setServletContext(arg0);//和当前 ServletContext 关联
        /**
         * 注册 Spring MVC 的 DispatcherServlet
         */
        Dynamic servlet =arg0.addServlet("dispatcher", new DispatcherServlet
                                        (ctx));
        servlet.addMapping("/");
        servlet.setLoadOnStartup(1);
        /**
         * 注册字符编码过滤器
         */
        javax.servlet.FilterRegistration.Dynamic filter =
                arg0.addFilter("characterEncodingFilter",
                        CharacterEncodingFilter.class);
        filter.setInitParameter("encoding", "UTF-8");
        filter.addMappingForUrlPatterns(null, false, "/*");
    }
}
```

9. 编写数据显示页面

在应用 ch7_1 的 src/main/webapp/WEB-INF/jsp/目录下创建数据显示页面 goodsList.jsp。具体代码如下:

```
<%@page language="java" contentType="text/html; charset=UTF-8" pageEncoding="UTF-8"%>
<%@taglib uri="http://java.sun.com/jsp/jstl/core" prefix="c" %>
<%
```

```jsp
    String path =request.getContextPath();
    String basePath = request.getScheme()+"://"+request.getServerName()+":"+
request.getServerPort()+path+"/";
%>
<!DOCTYPE html>
<html>
<head>
<base href="<%=basePath%>">
<meta charset="UTF-8">
<title>Insert title here</title>
<link rel="stylesheet" href="static/css/bootstrap.min.css" />
</head>
<body>
<div class="container">
        <div class="panel panel-primary">
            <div class="panel-heading">
                <h3 class="panel-title">商品列表</h3>
            </div>
            <div class="panel-body">
                <div class="table table-responsive">
                    <table class="table table-bordered table-hover">
                        <tbody class="text-center">
                            <tr>
                                <th>商品名称</th>
                                <th>商品详情</th>
                                <th>商品价格</th>
                                <th>创建日期</th>
                            </tr>
                            <c:forEach items="${goodsList }" var="goods">
                                <tr>
                                    <td>${goods.gname }</td>
                                    <td>${goods.gdescription }</td>
                                    <td>${goods.gprice }</td>
                                    <td>${goods.gdate }</td>
                                </tr>
                            </c:forEach>
                        </tbody>
                    </table>
                </div>
            </div>
        </div>
    </div>
</body>
</html>
```

10. 测试应用

发布应用 ch7_1 并启动 Tomcat 服务器后,通过地址 http://localhost:8080/ch7_1/goods/input 测试应用。如果数据输入有误,验证不通过时,回到信息输入页面,如图 7.3 所示。

图 7.3　信息输入错误时的页面

7.2.4　实践环节

参考 7.2.3 小节,创建应用 practice724,应用中有个输入页面,效果如图 7.4 所示。在该应用中创建一个实现 org.springframework.validation.Validator 接口的验证器类,验证器对输入页面的输入项进行验证。要求如下:

(1) 所有输入项不能为空。

图 7.4　实践环节的输入页面

(2) 年龄必须在 18 至 65 岁之间。

(3) E-mail 满足正常的格式。

7.3 JSR 303 验证

对于 JSR 303 验证，目前有两个实现，一个是 Hibernate Validator，另一个是 Apache BVal。本书采用的是 Hibernate Validator，注意它和 Hibernate 无关，只是使用它进行数据验证。

7.3.1 JSR 303 验证配置

1. 下载与安装 Hibernate Validator

读者如果不使用 Maven 添加 Hibernate Validator 依赖，可以通过地址 https://sourceforge.net/projects/hibernate/files/hibernate-validator/下载 Hibernate Validator。编写本书时，最新版是 hibernate-validator-6.1.2.Final-dist.zip。首先，解压缩下载的压缩包；然后将\hibernate-validator-6.1.2.Final\dist 目录下的 hibernate-validator-6.1.2.Final.jar 和\hibernate-validator-6.1.2.Final\dist\lib\required 目录下的 jakarta.validation-api-2.0.2.jar、classmate-1.3.4.jar、jboss-logging-3.3.2.Final.jar 复制到应用的\WEB-INF\lib 目录下。

2. Maven 安装 Hibernate Validator

读者欲使用 Maven 下载 Hibernate Validator 时，可以在应用的 pom.xml 中，添加如下依赖：

```
<dependency>
    <groupId>org.hibernate.validator</groupId>
    <artifactId>hibernate-validator</artifactId>
    <version>6.1.2.Final</version>
</dependency>
```

7.3.2 标注类型

使用 Hibernate Validator 验证表单时，需要利用它的标注类型在实体模型的属性上嵌入约束。

1. 空检查

@Null：验证对象是否为 null。

@NotNull：验证对象是否不为 null，无法查检长度为 0 的字符串。

@NotBlank：检查约束字符串是不是 null，还有被 trim 后的长度是否大于 0，只针对字符串，且会去掉前后空格。

@NotEmpty：检查约束元素是否为 null 或者是 empty。

示例如下：

```
@NotBlank(message="{goods.gname.required}")//goods.gname.required 为属性文件
                                            //的错误代码
private String gname;
```

2. boolean 检查

@AssertTrue：验证 boolean 属性是否为 true。
@AssertFalse：验证 boolean 属性是否为 false。

示例如下：

```
@AssertTrue
private boolean isLogin;
```

3. 长度检查

@Size(min=，max=)：验证对象（Array，Collection，Map，String）长度是否在给定的范围之内。
@Length(min=，max=)：验证字符串长度是否在给定的范围之内。

示例如下：

```
@Length(min=1,max=100)
private String gdescription;
```

4. 日期检查

@Past：验证 Date 和 Calendar 对象是否在当前时间之前。
@Future：验证 Date 和 Calendar 对象是否在当前时间之后。
@Pattern：验证 String 对象是否符合正则表达式的规则。

示例如下：

```
@Past(message="{gdate.invalid}")
private Date gdate;
```

5. 数值检查

@Min：验证 Number 和 String 对象是否大于等于指定的值。
@Max：验证 Number 和 String 对象是否小于等于指定的值。
@DecimalMax：被标注的值必须不大于约束中指定的最大值，这个约束的参数是一个通过 BigDecimal 定义的最大值的字符串表示，小数存在精度。
@DecimalMin：被标注的值必须不小于约束中指定的最小值，这个约束的参数是一个通过 BigDecimal 定义的最小值的字符串表示，小数存在精度。

@Digits：验证 Number 和 String 的构成是否合法。

@Digits(integer＝,fraction＝)：验证字符串是否符合指定格式的数字，interger 指定整数精度，fraction 指定小数精度。

@Range(min＝,max＝)：检查数字是否介于 min 和 max 之间。

@Valid：对关联对象进行校验，如果关联对象是个集合或者数组，那么对其中的元素进行校验，如果是一个 map，则对其中的值部分进行校验。

@CreditCardNumber：信用卡验证。

@Email：验证是否是邮件地址，如果为 null，不进行验证，即通过验证。

示例如下：

```
@Range(min=0,max=100,message="{gprice.invalid}")
private double gprice;
```

7.3.3 验证示例

创建应用 ch7_2，该应用实现的功能与 7.2.3 小节 ch7_1 应用相同。

【例 7-2】 Hibernate Validator 验证示例。在应用 ch7_2 中不需要创建验证器类 GoodsValidator。另外，Service 层、View 层和 Web 的 Java 配置类都与 ch7_1 应用相同，直接复制到 ch7_2 中即可。与 ch7_1 应用不同的是 pom.xml、模型类、错误消息属性文件、控制器类和 Spring MVC 的 Java 配置类，具体实现如下。

1. pom.xml 文件

在 ch7_2 的 pom.xml 文件中除了添加 Spring MVC 和 JSTL 依赖外，还需要添加 Hibernate Validator 依赖。ch7_2 的 pom.xml 文件内容如下：

```xml
<project xmlns="http://maven.apache.org/POM/4.0.0"
    xmlns:xsi="http://www.w3.org/2001/XMLSchema-instance"
    xsi:schemaLocation="http://maven.apache.org/POM/4.0.0
https://maven.apache.org/xsd/maven-4.0.0.xsd">
    <modelVersion>4.0.0</modelVersion>
    <groupId>com.cn.chenheng</groupId>
    <artifactId>ch7_2</artifactId>
    <version>0.0.1-SNAPSHOT</version>
    <packaging>war</packaging>
    <properties>
        <!--spring 版本号 -->
        <spring.version>5.2.3.RELEASE</spring.version>
    </properties>
    <dependencies>
        <dependency>
            <groupId>org.springframework</groupId>
            <artifactId>spring-webmvc</artifactId>
```

```xml
            <version>${spring.version}</version>
        </dependency>
        <dependency>
            <groupId>javax.servlet</groupId>
            <artifactId>jstl</artifactId>
            <version>1.2</version>
        </dependency>
        <dependency>
            <groupId>org.hibernate.validator</groupId>
            <artifactId>hibernate-validator</artifactId>
            <version>6.1.2.Final</version>
        </dependency>
    </dependencies>
</project>
```

2. 模型类

在模型类 Goods 中,利用 Hibernate Validator 的标注类型对属性进行验证,具体代码如下:

```java
package pojo;
import java.util.Date;
import javax.validation.constraints.Past;
import javax.validation.constraints.NotBlank;
import org.hibernate.validator.constraints.Range;
import org.springframework.format.annotation.DateTimeFormat;
public class Goods {
    //goods.gname.required错误消息key
    @NotBlank(message="{goods.gname.required}")
    private String gname;
    @NotBlank(message="{goods.gdescription.required}")
    private String gdescription;
    @Range(min=0,max=100,message="{gprice.invalid}")
    private double gprice;
    //日期格式化
    @DateTimeFormat(pattern="yyyy-MM-dd")
    @Past(message="{gdate.invalid}")
    private Date gdate;
    //省略 setter 和 getter 方法
}
```

3. 错误消息属性文件

将 ch7_1 的错误消息属性文件复制到 ch7_2 的 src/main/resources 目录下,并将名称修改为 ValidationMessages.properties。这是因为 Hibernate Validator 验证时，默认从 classpath 目录的 ValidationMessages.properties 文件中获取错误消息。

4. 控制器类

在控制器类 GoodsController 中，使用@Valid 注解开启对模型对象验证，具体代码如下：

```java
package controller;
import javax.validation.Valid;
import org.apache.commons.logging.Log;
import org.apache.commons.logging.LogFactory;
import org.springframework.beans.factory.annotation.Autowired;
import org.springframework.stereotype.Controller;
import org.springframework.ui.Model;
import org.springframework.validation.BindingResult;
import org.springframework.web.bind.annotation.ModelAttribute;
import org.springframework.web.bind.annotation.RequestMapping;
import pojo.Goods;
import service.GoodsService;
@Controller
@RequestMapping("/goods")
public class GoodsController {
    private static final Log logger = LogFactory.getLog(GoodsController.class);
    @Autowired
    private GoodsService goodsService;
    @RequestMapping("/input")
    public String input(Model model){
        model.addAttribute("goods", new Goods());
        return "addGoods";
    }
    /**
     * 使用@Valid 开启对模型对象进行验证
     * BindingResult 参数一定紧跟在@Valid 注解方法的后面，否则在校验不通过时 Spring 将直接抛出异常
     */
    @RequestMapping("/save")
    public String save (Model model, @ModelAttribute @Valid Goods goods,
                       BindingResult result){
        if(result.hasErrors()){
            return "addGoods";
        }
        goodsService.save(goods);
        logger.info("添加成功");
```

```
        model.addAttribute("goodsList", goodsService.getGoods());
        return "goodsList";
    }
}
```

5. Spring MVC 的 Java 配置类

在应用 ch7_2 的 Spring MVC 的 Java 配置类 SpringMVCConfig 中,只需配置视图解析器和静态资源,而错误消息属性文件不再需要配置(使用了默认目录和文件名)。Java 配置类 SpringMVCConfig 的代码如下:

```
package config;
import org.springframework.context.annotation.Bean;
import org.springframework.context.annotation.ComponentScan;
import org.springframework.context.annotation.Configuration;
import org.springframework.web.servlet.config.annotation.EnableWebMvc;
import org.springframework.web.servlet.config.annotation.ResourceHandlerRegistry;
import org.springframework.web.servlet.config.annotation.WebMvcConfigurer;
import org.springframework.web.servlet.view.InternalResourceViewResolver;
@Configuration
@EnableWebMvc // 开启 Spring MVC 的支持
@ComponentScan(basePackages = { "controller", "service"}) // 扫描基本包
public class SpringMVCConfig implements WebMvcConfigurer {
    /**
     * 配置视图解析器
     */
    @Bean
    public InternalResourceViewResolver getViewResolver() {
        InternalResourceViewResolver viewResolver = new
        InternalResourceViewResolver();
        viewResolver.setPrefix("/WEB-INF/jsp/");
        viewResolver.setSuffix(".jsp");
        return viewResolver;
    }
    /**
     * 配置静态资源(不需要 DispatcherServlet 转发的请求)
     */
    @Override
    public void addResourceHandlers(ResourceHandlerRegistry registry) {
        registry.addResourceHandler("/static/**").addResourceLocations
        ("/static/");
    }
}
```

6. 测试应用

发布应用 ch7_2 到服务器上后,通过地址 http://localhost:8080/ch7_2/goods/input 测试应用 ch7_2。

7.3.4 实践环节

参考 7.3.3 小节,创建应用 practice734,在该应用中使用 Hibernate Validator 对输入页面的输入项进行验证。要求如下:

(1) 用户名不能为空,并且长度在 3 至 10。
(2) 使用正则表达式验证手机号。
(3) 生日满足日期格式,并且是在 1990-01-01 至 2020-07-31。
(4) 输入页面的运行效果如图 7.5 所示。

图 7.5 输入页面

7.4 本章小结

本章重点讲解了 Spring 验证的编写和 JSR 303 验证的使用方法。不管哪种验证方式,都需要注意验证流程。

习 题 7

1. 如何创建 Spring 验证器类?
2. 举例说明 Hibernate Validator 验证的标注类型的使用方法。

国 际 化

学习目的与要求

本章重点讲解 Spring MVC 国际化的实现方法。通过本章的学习,读者应理解 Spring MVC 国际化的设计思想,掌握 Spring MVC 国际化的实现方法。

本章主要内容

- Java 国际化的思想。
- Spring MVC 的国际化。
- 用户自定义切换语言。

国际化是商业软件系统的一个基本要求,因为当今的软件系统需要面对全球的浏览者。国际化的目的,就是根据用户的语言环境不同输出与之相应的页面给用户,以示友好。

Spring MVC 的国际化主要有页面信息国际化以及错误消息国际化。错误消息在"第 7 章 数据验证"已讲解,本章主要介绍如何在页面输出国际化消息。最后,本章将示范一个让用户自行选择语言的示例。

8.1 程序国际化概述

程序国际化已成为Web应用的基本要求。随着网络的发展,大部分Web站点面对的已经不再是本地或者本国的浏览者,而是来自全世界各国各地区的浏览者,因此国际化成了Web应用不可或缺的一部分。

8.1.1 Java国际化的思想

Java国际化的思想是将程序中的信息放在资源文件中,程序根据支持的国家及语言环境读取相应的资源文件。资源文件是key-value对,每个资源文件中的key是不变的,但value则随不同国家/语言变化。

在Java程序的国际化中,主要通过两个类来完成:

(1) java.util.Locale:用于提供本地信息,通常称它为语言环境。不同的语言、不同的国家和地区采用不同的Locale对象来表示。

(2) java.util.ResourceBundle:该类称为资源包,包含了特定于语言环境的资源对象。当程序需要一个特定于语言环境的资源时(如字符串资源),程序可以从适合当前用户语言环境的资源包中加载它。采用这种方式,可以编写独立于用户语言环境的程序代码,而与特定语言环境相关的信息则通过资源包来提供。

为了实现Java程序的国际化,必须事先提供程序需要的资源文件。资源文件的内容由很多key-value对组成,其中key是程序使用的部分,而value则是程序界面的显示。

资源文件的命名可以有如下3种形式:

(1) baseName.properties。

(2) baseName_language.properties。

(3) baseName_language_country.properties。

baseName是资源文件的基本名称,由用户自由定义。但是language和country就必须为Java所支持的语言和国家/地区代码。例如:

中国大陆:baseName_zh_CN.properties

美国:baseName_en_US.properties

Java中的资源文件只支持ISO-8859-1编码格式字符,直接编写中文会出现乱码。可以使用Java命令native2ascii.exe解决资源文件的中文乱码问题。使用STS编写资源文件,在保存资源文件时,STS将自动执行native2ascii.exe命令。因此,在STS中的资源文件不会出现中文乱码问题。

8.1.2 Java支持的语言和国家

java.util.Locale类的常用构造方法如下:

(1) public Locale(String language)。

(2) public Locale(Stringlanguage,String country)。

其中language表示语言,它的取值是由小写的两个字母组成的语言代码。country

表示国家或地区，它的取值是由大写的两个字母组成的国家或地区代码。

实际上，Java 并不能支持所有国家和语言，如果需要获取 Java 所支持的语言和国家，可以通过调用 Locale 类的 getAvailableLocales 方法获取，该方法返回一个 Locale 数组，该数组里包含了 Java 所支持的语言和国家。

下面的 Java 程序简单示范了如何获取 Java 所支持的国家和语言：

```java
package ch8_1;
import java.util.Locale;
public class Test {
    public static void main(String[] args) {
        // 返回 Java 所支持的语言和国家的数组
        Locale locales[] = Locale.getAvailableLocales();
        // 遍历数组元素，依次获取所支持的国家和语言
        for (int i = 0; i < locales.length; i++) {
            // 打印出所支持的国家和语言
            System.out.println(locales[i].getDisplayCountry() + "="
                + locales[i].getCountry() + " "
                + locales[i].getDisplayLanguage() + "="
                + locales[i].getLanguage());
        }
    }
}
```

8.1.3　Java 程序国际化

假设有如下的简单 Java 程序：

```java
package ch8_1;
public class TestI18N {
    public static void main(String[] args) {
        System.out.println("我要向不同国家的人民问好：您好！");
    }
}
```

为了让该程序支持国际化，需要将"我要向不同国家的人民问好：您好！"对应不同语言环境的字符串定义在不同的资源文件中。

首先，在 Web 应用的 src/main/resources 目录下新建资源文件 messageResource_zh_CN.properties 和 messageResource_en_US.properties。

其次，给资源文件 messageResource_zh_CN.properties 添加"hello＝我要向不同国家的人民问好：您好！"内容，保存后可看到图 8.1 所示的效果。

图 8.1 显示的内容看似是很多乱码，实际是 Unicode 编码文件内容。至此，资源文件 messageResource_zh_CN.properties 创建完成。

最后，给资源文件 messageResource_en_US.properties 添加"hello＝I want to say

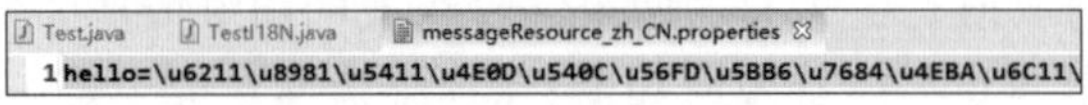
Test.java
TestI18N.java
messageResource_zh_CN.properties
1 hello=\u6211\u8981\u5411\u4E0D\u540C\u56FD\u5BB6\u7684\u4EBA\u6C11\

图 8.1 Unicode 编码资源文件

hello to all world!”内容。

现在将 TestI18N.java 程序修改成如下形式：

```
package ch8_1;
import java.util.Locale;
import java.util.ResourceBundle;
public class TestI18N {
    public static void main(String[] args) {
        //取得系统默认的国家语言环境
        Locale lc =Locale.getDefault();
        //根据国家语言环境加载资源文件,默认从 src/main/resources 目录下获取资源文件
        ResourceBundle rb =ResourceBundle.getBundle("messageResource", lc);
        //打印从资源文件中取得的信息
        System.out.println(rb.getString("hello"));
    }
}
```

上面程序中的打印语句打印的内容是从资源文件中读取的信息。如果在中文环境下运行程序，将打印“我要向不同国家的人民问好：您好！”；如果在“控制面板”中将机器的语言环境设置成美国，然后再次运行该程序，将打印“I want to say hello to all world!”。

需要注意的是，如果程序找不到对应国家/语言的资源文件，系统该怎么办？以简体中文环境为例，先搜索如下文件：

messageResource_zh_CN.properties

如果没有找到国家/语言匹配的资源文件，再搜索语言匹配文件，即搜索如下文件：

messageResource_zh.properties

如果上面的文件还没有搜索到，则搜索 baseName 匹配的文件，即搜索如下文件：

messageResource.properties

如果上面 3 个文件都找不到，则系统出现异常。

8.1.4 带占位符的国际化信息

在资源文件中的消息文本可以带参数，例如：

welcome={0}，欢迎学习 Spring MVC。

花括号中的数字是一个占位符，可以被动态的数据替换。在消息文本中的占位符可以使用 0 到 9 的数字，也就是说，消息文本的参数最多可以有 10 个。例如：

welcome={0}，欢迎学习 Spring MVC，今天是星期{1}。

要替换消息文本中的占位符，可以使用 java.text.MessageFormat 类，该类提供了一个静态方法 format()，用来格式化带参数的文本，format()方法定义如下：

```
public static String format(Stringpattern,Object ...arguments)
```

其中,pattern字符串就是一个带占位符的字符串,消息文本中的数字占位符将按照方法的参数顺序(从第二个参数开始)被替换。

替换占位符的示例代码如下:

```
package ch8_1;
import java.text.MessageFormat;
import java.util.Locale;
import java.util.ResourceBundle;
public class TestFormat {
    public static void main(String[] args) {
        //取得系统默认的国家语言环境
        Locale lc =Locale.getDefault();
        //根据国家语言环境加载资源文件
        ResourceBundle rb =ResourceBundle.getBundle("messageResource", lc);
        //从资源文件中取得的信息
        String msg =rb.getString("welcome");
        //替换消息文本中的占位符,消息文本中的数字占位符将按照参数的顺序
        //(从第二个参数开始)而被替换,即"我"替换{0},"5"替换{1}
        String msgFor =MessageFormat.format(msg, "我","5");
        System.out.println(msgFor);
    }
}
```

8.1.5 实践环节

编写一个Java应用程序,该程序从资源文件messageResource_zh_CN.properties中读取消息文本"practice815＝今天{0}很高兴,{1}也不错,明天就是星期{2}了。"。对应的英文资源文件是messageResource_en_US.properties,消息文本是"practice815＝ Today,{0} is very glad,{1} is too good,tomorrow will be {2}."。

8.2 Spring MVC 的国际化

Spring MVC 的国际化是建立在 Java 国际化的基础之上。Spring MVC 框架的底层国际化与 Java 国际化是一致的。作为一个良好的 MVC 框架,Spring MVC 将 Java 国际化的功能进行了封装和简化,开发者使用起来更加简单快捷。

由8.1节可知国际化和本地化应用程序时,需要具备以下两个条件:
(1) 将文本信息放到资源属性文件中。
(2) 选择和读取正确位置的资源属性文件。
下面讲解第二个条件的实现。

8.2.1　Spring MVC 加载资源属性文件

在 Spring MVC 中，不是直接使用 ResourceBundle 加载资源属性文件，而是利用 Bean（messageSource）告知 Spring MVC 框架从哪里获得资源属性文件。示例代码如下：

```
/**
 * 配置资源属性文件 messages.properties
 * Bean 的名字必须是 messageSource,这是 Spring 规定的
 */
@Bean("messageSource")
public ResourceBundleMessageSource getMessageSource() {
       ResourceBundleMessageSource rbms =new ResourceBundleMessageSource();
       //默认从 src/main/resources 目录下查找资源属性文件
       rbms.setBasename("messages");
       rbms.setDefaultEncoding("UTF-8");
       return rbms;
}
```

上述 Bean 配置的是国际化资源文件的路径，setBasename()方法默认指定 classpath 目录下的 messages_zh_CN.properties 和 messages_en_US.properties 资源文件。当然也可以将国际化资源文件放在其他的路径下，如 /WEB-INF/resource/messages。

另外，名为 messageSource 的 Bean 是由 ResourceBundleMessageSource 实现的，并且 Bean 的名字必须是 messageSource。如果修改了国际化资源文件，需要重启 JVM。

最后，还需注意，如果有一组属性文件，则用 setBasenames()指定资源文件，示例代码如下：

```
@Bean("messageSource")
public ResourceBundleMessageSource getMessageSource() {
    ResourceBundleMessageSource rbms =new ResourceBundleMessageSource();
    String messageFiles[] ={"messages", "labels"};
    rbms.setBasenames(messageFiles);
    rbms.setDefaultEncoding("UTF-8");
    return rbms;
}
```

8.2.2　语言区域的选择

在 Spring MVC 中，可以使用语言区域解析器 Bean 选择语言区域。该 Bean 有 3 个常见实现：AcceptHeaderLocaleResolver、SessionLocaleResolver 以及 CookieLocaleResolver。

1. AcceptHeaderLocaleResolver

根据浏览器 HTTP Header 中的 accept-language 域判定（accept-language 域中一般包含了当前操作系统的语言设定，可通过 HttpServletRequest.getLocale 方法获得此域的内容）。

改变 Locale 是不支持的，即不能调用 LocaleResolver 接口的 setLocale（HttpServletRequest request，HttpServletResponse response，Locale locale）方法设置 Locale。

2. SessionLocaleResolver

根据用户本次会话过程中的语言设定决定语言区域（如：用户进入首页时选择语言种类，则此次会话周期内统一使用该语言设定）。

3. CookieLocaleResolver

根据 Cookie 判定用户的语言设定（Cookie 中保存着用户前一次的语言设定参数）。

由上述分析可知，SessionLocaleResolver 实现比较方便用户选择喜欢的语言种类，本章将使用该方法进行国际化实现。

使用 SessionLocaleResolver 实现的 Bean 定义如下：

```
/**
 * 配置 SessionLocaleResolver 实现的 Bean
 */
@Bean("localeResolver")
public LocaleResolver getLocaleResolver() {
    SessionLocaleResolver sessionLocaleResolver = new SessionLocaleResolver();
    Locale loc = new Locale("zh", "CN");
    //设置默认语言和国家
    sessionLocaleResolver.setDefaultLocale(loc);
    return sessionLocaleResolver;
}
```

如果基于 SessionLocaleResolver 和 CookieLocaleResolver 实现国际化，必须配置 LocaleChangeInterceptor 拦截器，示例代码如下：

```
/**
 * 配置 LocaleChangeInterceptor 拦截器
 */
@Override
public void addInterceptors(InterceptorRegistry registry) {
    LocaleChangeInterceptor localeChangeInterceptor = new LocaleChangeInterceptor();
    registry.addInterceptor(localeChangeInterceptor).addPathPatterns("/**");
}
```

8.2.3 使用 message 标签显示国际化信息

在 Spring MVC 框架中，可以使用 Spring 的 message 标签在 JSP 页面中显示国际化消息。使用 message 标签时，需要在 JSP 页面最前面使用 taglib 指令声明 spring 标签，代码如下：

```
<%@taglib prefix="spring" uri="http://www.springframework.org/tags" %>
```

message 标签有如下常用属性：
- code：获得国际化消息的 key。
- arguments：代表该标签的参数。如替换消息中的占位符，示例代码如下：

`<spring:message code="third" arguments="888,999" />`，third 对应的消息有两个占位符{0}和{1}

- argumentSeparator：用来分隔该标签参数的字符，默认为逗号。
- text：如果 code 属性不存在，或指定的 key 无法获取消息时，所显示的默认文本信息。

8.3 用户自定义切换语言示例

在许多成熟的商业软件系统中，可以让用户自由切换语言，而不是修改浏览器的语言设置。一旦用户选择了自己需要使用的语言环境，整个系统的语言环境将一直是这种语言环境。Spring MVC 也可以允许用户自行选择语言。本章通过 Web 应用 ch8_1 演示用户自定义切换语言，在该应用中使用 SessionLocaleResolver 实现国际化。

【例 8-1】 Spring MVC 实现用户自定义切换语言。

具体实现步骤如下。

1. 创建 Maven 项目并添加依赖的 JAR 包

创建一个名为 ch8_1 的 Maven 项目，并通过 pom.xml 文件添加项目的相关依赖 spring-webmvc。pom.xml 文件的内容如下：

```xml
<project xmlns="http://maven.apache.org/POM/4.0.0"
xmlns:xsi="http://www.w3.org/2001/XMLSchema-instance"
xsi:schemaLocation="http://maven.apache.org/POM/4.0.0
https://maven.apache.org/xsd/maven-4.0.0.xsd">
  <modelVersion>4.0.0</modelVersion>
  <groupId>com.cn.chenheng</groupId>
  <artifactId>ch8_1</artifactId>
  <version>0.0.1-SNAPSHOT</version>
  <packaging>war</packaging>
  <properties>
     <!--spring版本号 -->
     <spring.version>5.2.3.RELEASE</spring.version>
  </properties>
  <dependencies>
     <dependency>
        <groupId>org.springframework</groupId>
```

```xml
            <artifactId>spring-webmvc</artifactId>
            <version>${spring.version}</version>
        </dependency>
    </dependencies>
</project>
```

2. 创建国际化资源文件

在 src/main/resources 目录下创建中英文资源文件 messages_en_US.properties 和 messages_zh_CN.properties。

messages_en_US.properties 的内容如下：

```
first=first
second=second
third={0} third {1}
language.en=English
language.cn=Chinese
```

messages_zh_CN.properties 的内容如下：

```
first=\u7B2C\u4E00\u9875
second=\u7B2C\u4E8C\u9875
third={0} \u7B2C\u4E09\u9875 {1}
language.cn=\u4E2D\u6587
language.en=\u82F1\u6587
```

3. 创建视图 JSP 文件

在 src/main/webapp/WEB-INF/jsp/ 目录下创建 3 个 JSP 文件：first.jsp、second.jsp 和 third.jsp。

first.jsp 的代码如下：

```jsp
<%@page language="java" contentType="text/html; charset=UTF-8" pageEncoding="UTF-8"%>
<%@taglib prefix="spring" uri="http://www.springframework.org/tags" %>
<%
String path = request.getContextPath();
String basePath = request.getScheme()+"://"+request.getServerName()+":"+request.getServerPort()+path+"/";
%>
<!DOCTYPE html>
<html>
<head>
<base href="<%=basePath%>">
<meta charset="UTF-8">
```

```
<title>Insert title here</title>
</head>
<body>
    <a href="i18nTest?locale=zh_CN"><spring:message code="language.cn" />
</a>--
    <a href="i18nTest?locale=en_US"><spring:message code="language.en" />
</a>
    <br><br>
    <spring:message code="first"/><br><br>
    <a href="my/second"><spring:message code="second"/></a>
</body>
</html>
```

second.jsp 的代码如下:

```
<%@page language="java" contentType="text/html; charset=UTF-8" pageEncoding="UTF-8"%>
<%@taglib prefix="spring" uri="http://www.springframework.org/tags" %>
<%
String path = request.getContextPath();
String basePath = request.getScheme()+"://"+request.getServerName()+":"+request.getServerPort()+path+"/";
%>
<!DOCTYPE html>
<html>
<head>
<base href="<%=basePath%>">
<meta charset="UTF-8">
<title>Insert title here</title>
</head>
<body>
    <spring:message code="second"/><br><br>
    <a href="my/third"><spring:message code="third" arguments="888,999" /></a>
</body>
</html>
```

third.jsp 的代码如下:

```
<%@page language="java" contentType="text/html; charset=UTF-8" pageEncoding="UTF-8"%>
<%@taglib prefix="spring" uri="http://www.springframework.org/tags" %>
<%
String path = request.getContextPath();
String basePath = request.getScheme()+"://"+request.getServerName()+":"+request.getServerPort()+path+"/";
%>
```

```
<!DOCTYPE html>
<html>
<head>
<base href="<%=basePath%>">
<meta charset="UTF-8">
<title>Insert title here</title>
</head>
<body>
    <spring:message code="third" arguments="888,999" /><br><br>
    <a href="my/first"><spring:message code="first"/></a>
</body>
</html>
```

4. 创建控制器类

该应用有两个控制器类,一个是 I18NTestController,处理语言种类选择请求;另一个是 MyController,进行页面导航。在 src/main/java 目录下创建一个名为 controller 的包,并在该包中创建这两个控制器类。

I18NTestController.java 的代码如下:

```
package controller;
import java.util.Locale;
import org.springframework.stereotype.Controller;
import org.springframework.web.bind.annotation.RequestMapping;
@Controller
public class I18NTestController {
    @RequestMapping("/i18nTest")
    /**
     * locale 接收请求参数 locale 值,并存储到 session 中
     */
    public String first(Locale locale){
        return "first";
    }
}
```

MyController 的代码如下:

```
package controller;
import org.springframework.stereotype.Controller;
import org.springframework.web.bind.annotation.RequestMapping;
@Controller
@RequestMapping("/my")
public class MyController {
    @RequestMapping("/first")
    public String first(){
```

```
        return "first";
    }
    @RequestMapping("/second")
    public String second(){
        return "second";
    }
    @RequestMapping("/third")
    public String third(){
        return "third";
    }
}
```

5. 创建 Spring MVC 配置类和 Web 配置类

在 src/main/java 目录下创建一个名为 config 的包，并在该包中创建 Web 配置类 WebConfig 和 Spring MVC 配置类 SpringMVCConfig。

在 Web 配置类 WebConfig 中注册 Spring MVC 的 Java 配置类 SpringMVCConfig、Spring MVC 的 DispatcherServlet 及字符编码过滤器。WebConfig 的代码如下：

```
package config;
import javax.servlet.ServletContext;
import javax.servlet.ServletException;
import javax.servlet.ServletRegistration.Dynamic;
import org.springframework.web.WebApplicationInitializer;
import org.springframework.web.context.support.
        AnnotationConfigWebApplicationContext;
import org.springframework.web.filter.CharacterEncodingFilter;
import org.springframework.web.servlet.DispatcherServlet;
public class WebConfig implements WebApplicationInitializer{
    @Override
    public void onStartup(ServletContext arg0) throws ServletException {
        AnnotationConfigWebApplicationContext ctx
        =new AnnotationConfigWebApplicationContext();
        ctx.register(SpringMVCConfig.class);//注册 Spring MVC 的 Java 配置类
                                            //SpringMVCConfig
        ctx.setServletContext(arg0);//和当前 ServletContext 关联
        /**
         * 注册 Spring MVC 的 DispatcherServlet
         */
        Dynamic servlet =arg0.addServlet("dispatcher", new DispatcherServlet
(ctx));
        servlet.addMapping("/");
        servlet.setLoadOnStartup(1);
        /**
         * 注册字符编码过滤器
```

```
        */
        javax.servlet.FilterRegistration.Dynamic filter =
                arg0.addFilter("characterEncodingFilter",
                CharacterEncodingFilter.class);
        filter.setInitParameter("encoding", "UTF-8");
        filter.addMappingForUrlPatterns(null, false, "/*");
    }
}
```

在 Spring MVC 配置类 SpringMVCConfig 中配置视图解析器、消息属性文件、SessionLocaleResolver 实现的 Bean、LocaleChangeInterceptor 拦截器及静态资源。具体代码如下：

```
package config;
import java.util.Locale;
import org.springframework.context.annotation.Bean;
import org.springframework.context.annotation.ComponentScan;
import org.springframework.context.annotation.Configuration;
import org.springframework.context.support.ResourceBundleMessageSource;
import org.springframework.web.servlet.LocaleResolver;
import org.springframework.web.servlet.config.annotation.EnableWebMvc;
import org.springframework.web.servlet.config.annotation.InterceptorRegistry;
import org.springframework.web.servlet.config.annotation.ResourceHandlerRegistry;
import org.springframework.web.servlet.config.annotation.WebMvcConfigurer;
import org.springframework.web.servlet.i18n.LocaleChangeInterceptor;
import org.springframework.web.servlet.i18n.SessionLocaleResolver;
import org.springframework.web.servlet.view.InternalResourceViewResolver;
@Configuration
@EnableWebMvc // 开启 Spring MVC 的支持
@ComponentScan(basePackages = {"controller"}) // 扫描基本包
public class SpringMVCConfig implements WebMvcConfigurer {
    /**
     * 配置视图解析器
     */
    @Bean
    public InternalResourceViewResolver getViewResolver() {
        InternalResourceViewResolver viewResolver = new
        InternalResourceViewResolver();
        viewResolver.setPrefix("/WEB-INF/jsp/");
        viewResolver.setSuffix(".jsp");
        return viewResolver;
    }
    /**
     * 配置消息属性文件 messages.properties
     * Bean 的名字必须是 messageSource,这是 Spring 规定的
     */
```

```
@Bean("messageSource")
public ResourceBundleMessageSource getMessageSource() {
    ResourceBundleMessageSource rbms =new ResourceBundleMessageSource();
    rbms.setBasename("messages");
    rbms.setDefaultEncoding("UTF-8");
    return rbms;
}
/**
 * 配置 SessionLocaleResolver 实现的 Bean
 */
@Bean("localeResolver")
public LocaleResolver getLocaleResolver() {
    SessionLocaleResolver sessionLocaleResolver =new SessionLocaleResolver();
    Locale loc =new Locale("zh", "CN");
    //设置默认语言和国家
    sessionLocaleResolver.setDefaultLocale(loc);
    return sessionLocaleResolver;
}
/**
 * 配置 LocaleChangeInterceptor 拦截器
 */
@Override
public void addInterceptors(InterceptorRegistry registry) {
    LocaleChangeInterceptor localeChangeInterceptor =new
    LocaleChangeInterceptor();
    registry.addInterceptor(localeChangeInterceptor).addPathPatterns("/**");
}

/**
 * 配置静态资源(不需要 DispatcherServlet 转发的请求)
 */
@Override
public void addResourceHandlers(ResourceHandlerRegistry registry) {
    registry.addResourceHandler("/static/**").addResourceLocations
    ("/static/");
}
}
```

6. 发布应用并测试

首先，将应用 ch8_1 发布到 Tomcat 服务器，并启动 Tomcat 服务器。然后，通过地址 http://localhost:8080/ch8_1/my/first 测试第一个页面，运行结果如图 8.2 所示。

单击图 8.2 中的"第二页"超链接，打开 second.jsp 页面，运行结果如图 8.3 所示。

图 8.2　中文环境下 first.jsp 的运行效果 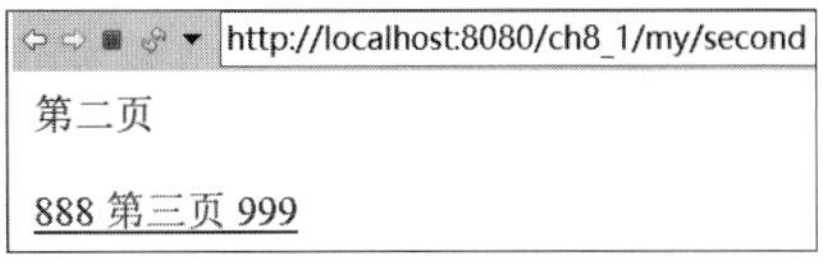　　图 8.3　中文环境下 second.jsp 的运行效果

单击图 8.3 中的"888 第三页 999"超链接,打开 third.jsp 页面,运行结果如图 8.4 所示。

单击图 8.2 中的"英文"超链接,打开英文环境下的 first.jsp 页面,运行结果如图 8.5 所示。

图 8.4　中文环境下 third.jsp 的运行效果　　图 8.5　英文环境下 first.jsp 的运行效果

单击图 8.5 中的"second"超链接,打开英文环境下的 second.jsp 页面,运行结果如图 8.6 所示。

单击图 8.6 中的"888third999"超链接,打开英文环境下的 third.jsp 页面,运行结果如图 8.7 所示。

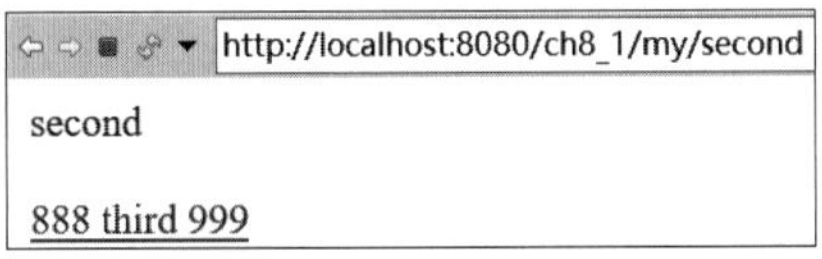

图 8.6　英文环境下 second.jsp 的运行效果　　图 8.7　英文环境下 third.jsp 的运行效果

8.4　本 章 小 结

本章主要讲解了 Spring MVC 的国际化知识,详细讲述了国际化资源文件的加载方式、语言区域选择、国际化信息显示。最后,本章给出了一个自行选择语言的示例,介绍了 Spring MVC 国际化的内在原理。

习　题　8

1. 在 JSP 页面中,可以通过 Spring 提供的(　　)标签来输出国际化信息。
　　A. input　　　　B. message　　　　C. submit　　　　D. text
2. 资源文件的后缀名为(　　)。
　　A. txt　　　　　B. doc　　　　　　C. property　　　 D. properties
3. 什么是国际化? 国际化资源文件的命名格式是什么?

统一异常处理

学习目的与要求

本章重点讲解如何使用 Spring MVC 框架进行统一异常处理。通过本章的学习，读者需要掌握 Spring MVC 框架统一异常处理的使用方法。

本章主要内容

- 简单异常处理 SimpleMappingExceptionResolver。
- 实现 HandlerExceptionResolver 接口自定义异常。
- 使用@ExceptionHandler 注解实现异常处理。

在 Spring MVC 应用的开发中，不管是对底层数据库操作，还是业务层操作，还是控制层操作，都不可避免地遇到各种可预知的、不可预知的异常需要处理。如果每个过程都单独处理异常，那么系统的代码耦合度高，工作量大且不好统一，以后维护的工作量也很大。

如果能将所有类型的异常处理从各层中解耦出来，既保证了相关处理过程的功能较单一，也实现了异常信息的统一处理和维护。幸运的是，Spring MVC 框架支持这样的实现。本章将从 SimpleMappingExceptionResolver、HandlerExceptionResolver、@ExceptionHandler 以及@ControllerAdvice 等方式讲解 Spring MVC 应用的统一异常处理。

9.1 示例介绍

为了验证 Spring MVC 框架异常处理的实际效果，需要开发一个测试应用 ch9_1，从 Dao 层、Service 层、Controller 层分别抛出不同的异常（SQLException、自定义异常和未知异常），然后分别集成 SimpleMappingExceptionResolver、HandlerExceptionResolver、@ExceptionHandler 以及 @ControllerAdvice 等方式进行异常处理，进而比较它们的优缺点。

四种异常处理方式的共通部分是 Dao 层、Service 层、View 层、MyException、TestExceptionController、Web 的 Java 配置类以及 pom.xml 文件。下面分别介绍这些共通部分。

1. 创建 Maven 项目并添加依赖的 JAR 包

创建一个名为 ch9_1 的 Maven 项目，并通过 pom.xml 文件添加项目的相关依赖 spring-webmvc。pom.xml 文件的内容如下：

```xml
<project xmlns="http://maven.apache.org/POM/4.0.0"
    xmlns:xsi="http://www.w3.org/2001/XMLSchema-instance"
    xsi:schemaLocation="http://maven.apache.org/POM/4.0.0
https://maven.apache.org/xsd/maven-4.0.0.xsd">
    <modelVersion>4.0.0</modelVersion>
    <groupId>com.cn.chenheng</groupId>
    <artifactId>ch9_1</artifactId>
    <version>0.0.1-SNAPSHOT</version>
    <packaging>war</packaging>
    <properties>
        <!--spring 版本号 -->
        <spring.version>5.2.3.RELEASE</spring.version>
    </properties>
    <dependencies>
        <dependency>
            <groupId>org.springframework</groupId>
            <artifactId>spring-webmvc</artifactId>
            <version>${spring.version}</version>
        </dependency>
    </dependencies>
</project>
```

2. 创建 Web 的 Java 配置类

在 src/main/java 目录下创建一个名为 config 的包，并在该包中创建 Web 的 Java 配置类 WebConfig。在该配置类中，注册 Spring MVC 的 Java 配置类以及 DispatcherServlet。

WebConfig 的代码如下：

```java
package config;
import javax.servlet.ServletContext;
import javax.servlet.ServletException;
import javax.servlet.ServletRegistration.Dynamic;
import org.springframework.web.WebApplicationInitializer;
import org.springframework.web.context.support.
AnnotationConfigWebApplicationContext;
import org.springframework.web.servlet.DispatcherServlet;
public class WebConfig implements WebApplicationInitializer{
    @Override
    public void onStartup(ServletContext arg0) throws ServletException {
        AnnotationConfigWebApplicationContext ctx
        =new AnnotationConfigWebApplicationContext();
        ctx.register(SpringMVCConfig.class);
        //注册 Spring MVC 的 Java 配置类 SpringMVCConfig
        ctx.setServletContext(arg0);//和当前 ServletContext 关联
        /**
         * 注册 Spring MVC 的 DispatcherServlet
         */
        Dynamic servlet = arg0.addServlet("dispatcher", new DispatcherServlet
                                        (ctx));
        servlet.addMapping("/");
        servlet.setLoadOnStartup(1);
    }
}
```

3. 创建自定义异常类

在 src/main/java 目录下创建一个名为 exception 的包，并在该包中创建自定义异常类 MyException。具体代码如下：

```java
package exception;
public class MyException extends Exception {
    private static final long serialVersionUID =1L;
    public MyException() {
        super();
    }
    public MyException(String message) {
        super(message);
    }
}
```

4. 创建 Dao 层

在 src/main/java 目录下创建一个名为 dao 的包,并在该包中创建 TestExceptionDao 类,该类中定义三个方法,分别抛出"数据库异常""自定义异常"以及"未知异常"。具体代码如下:

```java
package dao;
import java.sql.SQLException;
import org.springframework.stereotype.Repository;
import exception.MyException;
@Repository("testExceptionDao")
public class TestExceptionDao {
    public void daodb() throws Exception {
        throw new SQLException("Dao 中数据库异常");
    }
    public void daomy() throws Exception {
        throw new MyException("Dao 中自定义异常");
    }
    public void daono() throws Exception {
        throw new Exception("Dao 中未知异常");
    }
}
```

5. Service 层

在 src/main/java 目录下创建一个名为 service 的包,并在该包中创建 TestExceptionService 接口和 TestExceptionServiceImpl 实现类,该接口中定义六个方法,其中三个方法调用 Dao 层中的方法,三个是 Service 层的方法。Service 层的方法是为演示 Service 层的"数据库异常""自定义异常"以及"未知异常"而定义的。

TestExceptionService 的接口代码如下:

```java
package service;
public interface TestExceptionService {
    public void servicemy() throws Exception;
    public void servicedb() throws Exception;
    public void daomy() throws Exception;
    public void daodb() throws Exception;
    public void serviceno() throws Exception;
    public void daono() throws Exception;
}
```

TestExceptionServiceImpl 的实现类代码如下:

```java
package service;
import java.sql.SQLException;
```

```java
import org.springframework.beans.factory.annotation.Autowired;
import org.springframework.stereotype.Service;
import dao.TestExceptionDao;
import exception.MyException;
@Service("testExceptionService")
public class TestExceptionServiceImpl implements TestExceptionService{
    @Autowired
    private TestExceptionDao testExceptionDao;
    @Override
    public void servicemy() throws Exception {
        throw new MyException("Service中自定义异常");
    }
    @Override
    public void servicedb() throws Exception {
        throw new SQLException("Service中数据库异常");
    }
    @Override
    public void serviceno() throws Exception {
        throw new Exception("Service中未知异常");
    }
    @Override
    public void daomy() throws Exception {
        testExceptionDao.daomy();
    }
    @Override
    public void daodb() throws Exception {
        testExceptionDao.daodb();
    }
     public void daono() throws Exception{
        testExceptionDao.daono();
    }
}
```

6. 创建控制器类

在 src/main/java 目录下创建一个名为 controller 的包，并在该包中创建 TestExceptionController 控制器类，具体代码如下：

```java
package controller;
import java.sql.SQLException;
import org.springframework.beans.factory.annotation.Autowired;
import org.springframework.stereotype.Controller;
import org.springframework.web.bind.annotation.RequestMapping;
import exception.MyException;
```

```java
import service.TestExceptionService;
@Controller
public class TestExceptionController{
    @Autowired
    private TestExceptionService testExceptionService;
    @RequestMapping("/db")
    public void db() throws Exception {
        throw new SQLException("控制器中数据库异常");
    }
    @RequestMapping("/my")
    public void my() throws Exception {
        throw new MyException("控制器中自定义异常");
    }
    @RequestMapping("/no")
    public void no() throws Exception {
        throw new Exception("控制器中未知异常");
    }
    @RequestMapping("/servicedb")
    public void servicedb() throws Exception {
            testExceptionService.servicedb();;
    }
    @RequestMapping("/servicemy")
    public void servicemy() throws Exception {
            testExceptionService.servicemy();
    }
    @RequestMapping("/serviceno")
    public void serviceno() throws Exception {
            testExceptionService.serviceno();
    }
    @RequestMapping("/daodb")
    public void daodb() throws Exception {
            testExceptionService.daodb();
    }
    @RequestMapping("/daomy")
    public void daomy() throws Exception {
            testExceptionService.daomy();
    }
    @RequestMapping("/daono")
    public void daono() throws Exception {
            testExceptionService.daono();
    }
}
```

7. 创建 View 层

View 层共有 4 个 JSP 页面,下面分别介绍。

在 src/main/webapp 目录下创建应用程序的首页面 index.jsp。index.jsp 的代码如下:

```
<%@page language="java" contentType="text/html; charset=UTF-8" pageEncoding="UTF-8"%>
<%
String path =request.getContextPath();
String basePath = request.getScheme()+"://"+ request.getServerName()+":"+ request.getServerPort()+path+"/";
%>
<!DOCTYPE html>
<html>
<head>
<base href="<%=basePath%>">
<meta charset="UTF-8">
<title>Insert title here</title>
</head>
<body>
<h1>所有的演示例子</h1>
<h3><a href="daodb">处理 dao 中数据库异常</a></h3>
<h3><a href="daomy">处理 dao 中自定义异常</a></h3>
<h3><a href="daono">处理 dao 未知错误</a></h3>
<hr>
<h3><a href="servicedb">处理 service 中数据库异常</a></h3>
<h3><a href="servicemy">处理 service 中自定义异常</a></h3>
<h3><a href="serviceno">处理 service 未知错误</a></h3>
<hr>
<h3><a href="db">处理 controller 中数据库异常</a></h3>
<h3><a href="my">处理 controller 中自定义异常</a></h3>
<h3><a href="no">处理 controller 未知错误</a></h3>
</body>
</html>
```

在 src/main/webapp/WEB-INF 目录下创建一个名为 jsp 的文件夹。在该文件夹中创建未知异常对应页面 error.jsp(设置 isErrorPage="true",因为使用 exception 内置对象)。error.jsp 的代码如下:

```
<%@page language="java" contentType="text/html; charset=UTF-8" isErrorPage="true"
pageEncoding="UTF-8"%>
<%
```

```jsp
String path =request.getContextPath();
String basePath = request.getScheme()+"://"+request.getServerName()+":"+
request.getServerPort()+path+"/";
%>
<!DOCTYPE html>
<html>
<head>
<base href="<%=basePath%>">
<meta charset="UTF-8">
<title>Insert title here</title>
</head>
<body>
<%
    exception.printStackTrace(response.getWriter());
%>
<H1>未知错误:</H1><%=exception%>
</body>
</html>
```

在 jsp 的文件夹中创建自定义异常对应页面 my-error.jsp, 代码如下:

```jsp
<%@page language="java" contentType="text/html; charset=UTF-8" isErrorPage="true"
pageEncoding="UTF-8"%>
<%
String path =request.getContextPath();
String basePath = request.getScheme()+"://"+request.getServerName()+":"+
request.getServerPort()+path+"/";
%>
<!DOCTYPE html>
<html>
<head>
<base href="<%=basePath%>">
<meta charset="UTF-8">
<title>Insert title here</title>
</head>
<body>
<%
    exception.printStackTrace(response.getWriter());
%>
<H1>自定义异常错误:</H1><%=exception%>
</body>
</html>
```

在 jsp 的文件夹中创建 SQL 异常对应页面 sql-error.jsp, 代码如下:

```jsp
<%@page language="java" contentType="text/html; charset=UTF-8" isErrorPage="true"
pageEncoding="UTF-8"%>
<%
String path = request.getContextPath();
String basePath = request.getScheme()+"://"+request.getServerName()+":"+request.getServerPort()+path+"/";
%>
<!DOCTYPE html>
<html>
<head>
<base href="<%=basePath%>">
<meta charset="UTF-8">
<title>Insert title here</title>
</head>
<body>
<%
    exception.printStackTrace(response.getWriter());
%>
<H1>数据库异常错误:</H1><%=exception%>
</body>
</html>
```

9.2　SimpleMappingExceptionResolver 类

使用 org.springframework.web.servlet.handler.SimpleMappingExceptionResolver 类统一处理异常时，需要在 Spring MVC 的配置类中配置异常类和 View 的对应关系。Spring MVC 配置类 SpringMVCConfig 的代码如下：

```java
package config;
import java.util.Properties;
import org.springframework.context.annotation.Bean;
import org.springframework.context.annotation.ComponentScan;
import org.springframework.context.annotation.Configuration;
import org.springframework.web.servlet.config.annotation.EnableWebMvc;
import org.springframework.web.servlet.config.annotation.WebMvcConfigurer;
import org.springframework.web.servlet.handler.SimpleMappingExceptionResolver;
import org.springframework.web.servlet.view.InternalResourceViewResolver;
@Configuration
@EnableWebMvc  // 开启 Spring MVC 的支持
@ComponentScan(basePackages = {"controller", "dao", "service"})
public class SpringMVCConfig implements WebMvcConfigurer {
    /**
```

```java
 * 配置视图解析器
 */
@Bean
public InternalResourceViewResolver getViewResolver() {
    InternalResourceViewResolver viewResolver = new
    InternalResourceViewResolver();
    viewResolver.setPrefix("/WEB-INF/jsp/");
    viewResolver.setSuffix(".jsp");
    return viewResolver;
}
/**
 * 配置SimpleMappingExceptionResolver(异常类与View的对应关系)
 */
@Bean
public SimpleMappingExceptionResolver getExceptionResolver() {
    SimpleMappingExceptionResolver ser = new SimpleMappingExceptionResolver();
    //定义默认的异常处理页面
    ser.setDefaultErrorView("error");
    //定义需要特殊处理的异常,用类名或完全路径名作为key,异常页名作为值
    Properties mappings = new Properties();
    mappings.put("exception.MyException", "my-error");
    mappings.put("java.sql.SQLException", "sql-error");
    ser.setExceptionMappings(mappings);
    return ser;
}
}
```

配置完成后,即可通过SimpleMappingExceptionResolver异常处理类统一处理9.1节中的异常。

发布应用ch9_1到Tomcat服务器,通过地址http://localhost:8080/ch9_1/测试应用。

9.3　HandlerExceptionResolver 接口

org.springframework.web.servlet.HandlerExceptionResolver接口用于解析请求处理过程中产生的异常。开发者可以开发该接口的实现类,进行Spring MVC应用的异常统一处理。在ch9_1应用的exception包中创建一个HandlerExceptionResolver接口的实现类MyExceptionHandler。该类是一个组件类,在该类中重写resolveException接口方法,将异常与异常信息显示页面对应起来。具体代码如下:

```java
package exception;
import java.sql.SQLException;
import java.util.HashMap;
```

```java
import java.util.Map;
import javax.servlet.http.HttpServletRequest;
import javax.servlet.http.HttpServletResponse;
import org.springframework.stereotype.Component;
import org.springframework.web.servlet.HandlerExceptionResolver;
import org.springframework.web.servlet.ModelAndView;
@Component
public class MyExceptionHandler implements HandlerExceptionResolver {
    @Override
    public ModelAndView resolveException(HttpServletRequest arg0, HttpServletResponse
                                          arg1, Object arg2, Exception arg3) {
        Map<String, Object>model =new HashMap<String, Object>();
        // 根据不同错误转向不同页面(统一处理),即异常与view的对应关系
        if (arg3 instanceof MyException) {
            return new ModelAndView("my-error", model);
        } else if (arg3 instanceof SQLException) {
            return new ModelAndView("sql-error", model);
        } else {
            return new ModelAndView("error", model);
        }
    }
}
```

需要将实现类 MyExceptionHandler 的 Bean 托管给 Spring 容器,即在 Spring MVC 的配置类 SpringMVCConfig 中使用@ComponentScan 注解扫描进来。同时,需要将 9.2 节的 SimpleMappingExceptionResolver 的配置注释掉。SpringMVCConfig 的代码修改为:

```java
package config;
import org.springframework.context.annotation.Bean;
import org.springframework.context.annotation.ComponentScan;
import org.springframework.context.annotation.Configuration;
import org.springframework.web.servlet.config.annotation.EnableWebMvc;
import org.springframework.web.servlet.config.annotation.WebMvcConfigurer;
import org.springframework.web.servlet.view.InternalResourceViewResolver;
@Configuration
@EnableWebMvc // 开启 Spring MVC 的支持
@ComponentScan(basePackages ={ "controller", "dao", "service", "exception"})
public class SpringMVCConfig implements WebMvcConfigurer {
    /**
     * 配置视图解析器
     */
    @Bean
    public InternalResourceViewResolver getViewResolver() {
```

```
        InternalResourceViewResolver viewResolver =new 
        InternalResourceViewResolver();
        viewResolver.setPrefix("/WEB-INF/jsp/");
        viewResolver.setSuffix(".jsp");
        return viewResolver;
    }
}
```

配置完成后，发布应用 ch9_1 到 Tomcat 服务器，通过地址 http://localhost:8080/ch9_1/测试应用。

9.4 @ExceptionHandler 注解

如果在某个 Controller 中有一个使用@ExceptionHandler 注解修饰的方法，那么该 Controller 中任何方法抛出异常时，都由@ExceptionHandler 注解修饰的方法处理异常。

为了提高代码重用，首先创建所有控制器类的父类 BaseController 类；其次在该类中使用@ExceptionHandler 注解声明异常处理方法；最后处理异常的控制器类继承该父类即可。具体实现步骤如下。

首先，在 controller 包中创建一个名为 BaseController 的抽象类，并在该类中使用@ExceptionHandler 注解声明异常处理方法，具体代码如下：

```
package controller;
import java.sql.SQLException;
import org.springframework.web.bind.annotation.ExceptionHandler;
import exception.MyException;
public abstract class BaseController {
    /** 基于@ExceptionHandler 异常处理 */
    @ExceptionHandler
    public String exception(Exception ex)  {
        // 根据不同错误转向不同页面，即异常与 View 的对应关系
        if(ex instanceof SQLException) {
            return "sql-error";
        }else if(ex instanceof MyException) {
            return "my-error";
        } else {
            return "error";
        }
    }
}
```

然后，将所有需要异常处理的 Controller 都继承 BaseController 类，示例代码如下：

```
@Controller
public class TestExceptionController extends BaseController{
```

```
    ...
}
```

使用@ExceptionHandler 注解声明统一处理异常时,不需要配置任何信息。这时 ch9_1 应用的 SpringMVCConfig 的代码修改为:

```
package config;
import org.springframework.context.annotation.Bean;
import org.springframework.context.annotation.ComponentScan;
import org.springframework.context.annotation.Configuration;
import org.springframework.web.servlet.config.annotation.EnableWebMvc;
import org.springframework.web.servlet.config.annotation.WebMvcConfigurer;
import org.springframework.web.servlet.view.InternalResourceViewResolver;
@Configuration
@EnableWebMvc   //开启 Spring MVC 的支持
@ComponentScan(basePackages = { "controller", "dao", "service"})
public class SpringMVCConfig implements WebMvcConfigurer {
    /**
     * 配置视图解析器
     */
    @Bean
    public InternalResourceViewResolver getViewResolver() {
        InternalResourceViewResolver viewResolver =new
        InternalResourceViewResolver();
        viewResolver.setPrefix("/WEB-INF/jsp/");
        viewResolver.setSuffix(".jsp");
        return viewResolver;
    }
}
```

配置完成后,发布应用 ch9_1 到 Tomcat 服务器,通过地址 http://localhost:8080/ch9_1/测试应用。

9.5 @ControllerAdvice 注解

使用 9.4 节的父类 Controller 进行异常处理时,也有其自身的缺点,那就是代码耦合性太高。可以使用@ControllerAdvice 注解降低这种父子耦合关系。

@ControllerAdvice 注解,顾名思义,是一个增强的 Controller。使用该 Controller,可以实现三个方面的功能:全局异常处理、全局数据绑定以及全局数据预处理。本节将学习如何使用@ControllerAdvice 注解进行全局异常处理。

使用@ControllerAdvice 注解的类是当前 Spring MVC 应用中所有类的统一异常处理类,该类中使用@ExceptionHandler 注解的方法统一处理异常,不需要在每个 Controller 中逐一定义异常处理方法,这是因为对所有注解了@RequestMapping 的控器方法有效。

下面讲解如何使用@ControllerAdvice注解进行 ch9_1 的全局异常处理。

首先,在 ch9_1 的 controller 包中创建名为 GlobalExceptionHandlerController 的类。使用 @ControllerAdvice 注解修饰该类,并将 9.4 节父类 BaseController 中使用 @ExceptionHandler 注解修饰的方法移到该类中,具体代码如下:

```
package controller;
import java.sql.SQLException;
import org.springframework.web.bind.annotation.ControllerAdvice;
import org.springframework.web.bind.annotation.ExceptionHandler;
import exception.MyException;
@ControllerAdvice
public class GlobalExceptionHandlerController {
    /** 基于@ExceptionHandler 异常处理 */
    @ExceptionHandler
    public String exception(Exception ex)  {
        // 根据不同错误转向不同页面,即异常与 View 的对应关系
        if(ex instanceof SQLException) {
            return "sql-error";
        }else if(ex instanceof MyException) {
            return "my-error";
        } else {
            return "error";
        }
    }
}
```

其次,将 TestExceptionController 控制器类的继承关系去掉。

最后,发布应用 ch9_1 到 Tomcat 服务器,通过地址 http://localhost:8080/ch9_1/测试应用。

使用@ControllerAdvice 注解进行统一处理异常时,不需要配置任何信息。这时 ch9_1 应用的 SpringMVCConfig 的代码与 9.4 节相同。

9.6 本章小结

本章重点介绍了 Spring MVC 框架应用程序的统一异常处理的四种方法。从上面的处理过程可知,使用@ControllerAdvice 注解实现异常处理,具有集成简单、可扩展性好、不需要附加 Spring 配置等优点。

习 题 9

1. 简述 Spring MVC 框架中统一异常处理的常用方式。
2. 如何使用@ControllerAdvice 注解进行统一异常处理?

文件的上传和下载

学习目的与要求

本章重点讲解如何使用 Spring MVC 框架进行文件的上传与下载。通过本章的学习，读者需要掌握 Spring MVC 框架单文件上传、多文件上传以及文件下载。

本章主要内容

- 单文件上传。
- 多文件上传。
- 文件下载。

文件上传是 Web 应用经常需要面对的问题。对于 Java 应用而言，上传文件有多种方式：包括使用文件流手工编程上传、基于 commons-fileupload 组件的文件上传、基于 Servlet 3 及以上版本的文件上传等方式。本章将重点介绍如何使用 Spring MVC 框架进行文件上传。

10.1 文件上传

Spring MVC 框架的文件上传是基于 commons-fileupload 组件的文件上传,只不过 Spring MVC 框架在原有文件上传组件上做了进一步封装,简化了文件上传的代码实现,取消了不同上传组件上的编程差异。

10.1.1 commons-fileupload 组件

由于 Spring MVC 框架的文件上传是基于 commons-fileupload 组件的文件上传,因此,可通过 Maven 项目的 pom.xml 文件添加相关依赖(commons-fileupload 和 commons-io)。如果不使用 Maven 下载相关依赖,可将相关 JAR 包复制到 WEB-INF/lib 目录下。下面讲解如何下载这些 JAR 包。

Commons 是 Apache 开放源代码组织中的一个 Java 子项目,该项目包括文件上传、命令行处理、数据库连接池、XML 配置文件处理等模块。fileupload 就是其中用来处理基于表单的文件上传的子项目,commons-fileupload 组件性能优良,并支持任意大小文件的上传。

commons-fileupload 组件可以从 http://commons.apache.org/proper/commons-fileupload/ 上下载,本书采用的版本是 1.4。下载它的 Binaries 压缩包(commons-fileupload-1.4-bin.zip),解压后的目录中有两个子目录,分别是 lib 和 site。lib 目录下有个 JAR 文件:commons-fileupload-1.4.jar,该文件是 commons-fileupload 组件的类库。site 目录中是 commons-fileupload 组件的文档,也包括 API 文档。

commons-fileupload 组件依赖于 Apache 的另外一个项目:commons-io,该组件可以从 http://commons.apache.org/proper/commons-io/ 上下载,本书采用的版本是 2.6。下载它的 Binaries 压缩包(commons-io-2.6-bin.zip),解压缩后的目录中有 4 个 JAR 文件,其中有一个 commons-io-2.6.jar 文件,该文件是 commons-io 的类库。

10.1.2 基于表单的文件上传

标签:<input type="file"/>,在浏览器中会显示一个输入框和一个按钮,输入框可供用户填写本地文件的文件名和路径名,按钮可以让浏览器打开一个文件选择框,供用户选择文件。

文件上传的表单示例代码如下:

```
<form action="upload" method="post" enctype="multipart/form-data">
    <input type="file" name="myfile"/>
    ...
</form>
```

基于表单的文件上传,不要忘记使用 enctype 属性,并将它的值设置为 multipart/form-data。同时,表单的提交方式设置为 post。为什么需要这样呢?下面从 enctype 属性说起。

表单的 enctype 属性指定的是表单数据的编码方式，该属性有如下三个值：
- application/x-www-form-urlencoded：这是默认的编码方式，它只处理表单域里的 value 属性值。
- multipart/form-data：该编码方式以二进制流的方式处理表单数据，并将文件域指定文件的内容封装到请求参数里。
- text/plain：该编码方式当表单的 action 属性为 mailto:URL 的形式时才使用，主要适用于直接通过表单发送邮件的方式。

由上面三个属性的解释可知，基于表单上传文件时，enctype 的属性值应为 multipart/form-data。

10.1.3 MultipartFile 接口

在 Spring MVC 框架中，上传文件时，将文件相关信息及操作封装到 MultipartFile 对象中。因此，开发者只需要使用 MultipartFile 类型声明模型类的一个属性，即可以对被上传文件进行操作。该接口具有如下方法：
- byte[] getBytes()：以字节数组的形式返回文件的内容。
- String getContentType()：返回文件的内容类型。
- InputStream getInputStream()：返回一个 InputStream，从中读取文件的内容。
- String getName()：返回请求参数的名称。
- String getOriginalFilename()：返回客户端提交的原始文件名称。
- long getSize()：返回文件的大小，单位为字节。
- boolean isEmpty()：判断被上传文件是否为空。
- void transferTo(File destination)：将上传文件保存到目标目录下。

上传文件时，需要在配置类或配置文件中使用 Spring 的 org.springframework.web.multipart.commons.CommonsMultipartResolver 类配置用于文件上传的 Bean。

下面从单文件上传开始，讲解 MultipartFile 接口的使用方法。

10.1.4 单文件上传

本节通过一个应用案例 ch10_1 讲解 Spring MVC 框架如何实现单文件上传。

【例 10-1】Spring MVC 框架实现单文件上传。

具体实现步骤如下。

1. 创建 Maven 项目并添加依赖的 JAR 包

创建一个名为 ch10_1 的 Maven 项目，并通过 pom.xml 文件添加项目的相关依赖 spring-webmvc、commons-fileupload 和 jstl。pom.xml 文件的内容如下：

```
<project xmlns="http://maven.apache.org/POM/4.0.0"
    xmlns:xsi="http://www.w3.org/2001/XMLSchema-instance"
    xsi:schemaLocation="http://maven.apache.org/POM/4.0.0
    https://maven.apache.org/xsd/maven-4.0.0.xsd">
```

```xml
<modelVersion>4.0.0</modelVersion>
<groupId>com.cn.chenheng</groupId>
<artifactId>ch10_1</artifactId>
<version>0.0.1-SNAPSHOT</version>
<packaging>war</packaging>
<properties>
    <!--spring 版本号 -->
    <spring.version>5.2.3.RELEASE</spring.version>
</properties>
<dependencies>
    <dependency>
        <groupId>org.springframework</groupId>
        <artifactId>spring-webmvc</artifactId>
        <version>${spring.version}</version>
    </dependency>
    <dependency>
        <groupId>commons-fileupload</groupId>
        <artifactId>commons-fileupload</artifactId>
        <version>1.4</version>
    </dependency>
    <dependency>
        <groupId>javax.servlet</groupId>
        <artifactId>jstl</artifactId>
        <version>1.2</version>
    </dependency>
</dependencies>
</project>
```

2. 创建配置类

在 src/main/java 目录下创建一个名为 config 的包，并在该包中创建 Spring MVC 的配置类 SpringMVCConfig 和 Web 的配置类 WebConfig。

在 Spring MVC 的配置类 SpringMVCConfig 中配置视图解析器、静态资源以及上传文件的相关设置。SpringMVCConfig 的代码如下：

```java
package config;
import java.io.IOException;
import org.springframework.context.annotation.Bean;
import org.springframework.context.annotation.ComponentScan;
import org.springframework.context.annotation.Configuration;
import org.springframework.core.io.FileSystemResource;
import org.springframework.core.io.Resource;
import org.springframework.web.multipart.commons.CommonsMultipartResolver;
import org.springframework.web.servlet.config.annotation.EnableWebMvc;
```

```java
import org.springframework.web.servlet.config.annotation.
ResourceHandlerRegistry;
import org.springframework.web.servlet.config.annotation.WebMvcConfigurer;
import org.springframework.web.servlet.view.InternalResourceViewResolver;
@Configuration
@EnableWebMvc // 开启 Spring MVC 的支持
@ComponentScan(basePackages = { "controller"}) // 扫描基本包
public class SpringMVCConfig implements WebMvcConfigurer {
    /**
     * 配置视图解析器
     */
    @Bean
    public InternalResourceViewResolver getViewResolver() {
        InternalResourceViewResolver viewResolver = new
InternalResourceViewResolver();
        viewResolver.setPrefix("/WEB-INF/jsp/");
        viewResolver.setSuffix(".jsp");
        return viewResolver;
    }
    /**
     * 配置静态资源(不需要 DispatcherServlet 转发的请求)
     */
    @Override
    public void addResourceHandlers(ResourceHandlerRegistry registry) {
        registry.addResourceHandler("/static/**").addResourceLocations
("/static/");
    }
    /**
     * 配置上传文件的相关设置
     */
    @Bean("multipartResolver") //Bean 的名字固定
    public CommonsMultipartResolver getMultipartResolver() {
        CommonsMultipartResolver cmr = new CommonsMultipartResolver();
        //设置请求的编码格式,默认为 ISO-8859-1
        cmr.setDefaultEncoding("UTF-8");
        //设置允许上传文件的最大值,单位为字节
        cmr.setMaxUploadSize(5400000);
        //设置上传文件的临时路径
        //workspace\.metadata\.plugins\org.eclipse.wst.server.core\tmp0\
wtpwebapps\fileUpload
        Resource uploadTempDir = new FileSystemResource("fileUpload/temp");
        try {
            cmr.setUploadTempDir(uploadTempDir);
        } catch (IOException e) {
```

```java
            // TODO Auto-generated catch block
            e.printStackTrace();
        }
        return cmr;
    }
}
```

为防止中文乱码,需要在 Web 的配置类 WebConfig 中添加字符编码过滤器,WebConfig 的代码如下:

```java
package config;
import javax.servlet.ServletContext;
import javax.servlet.ServletException;
import javax.servlet.ServletRegistration.Dynamic;
import org.springframework.web.WebApplicationInitializer;
import org.springframework.web.context.support.
AnnotationConfigWebApplicationContext;
import org.springframework.web.filter.CharacterEncodingFilter;
import org.springframework.web.servlet.DispatcherServlet;
public class WebConfig implements WebApplicationInitializer{
    @Override
    public void onStartup(ServletContext arg0) throws ServletException {
        AnnotationConfigWebApplicationContext ctx
        =new AnnotationConfigWebApplicationContext();
        ctx.register(SpringMVCConfig.class);//注册 Spring MVC 的 Java 配置类
                                            //SpringMVCConfig
        ctx.setServletContext(arg0);//和当前 ServletContext 关联
        /**
         * 注册 Spring MVC 的 DispatcherServlet
         */
        Dynamic servlet =arg0.addServlet("dispatcher", new DispatcherServlet(ctx));
        servlet.addMapping("/");
        servlet.setLoadOnStartup(1);
        /**
         * 注册字符编码过滤器
         */
        javax.servlet.FilterRegistration.Dynamic filter =
                arg0.addFilter("characterEncodingFilter",
                CharacterEncodingFilter.class);
        filter.setInitParameter("encoding", "UTF-8");
        filter.addMappingForUrlPatterns(null, false, "/*");
    }
}
```

3. 创建文件选择页面

在 src/main/webapp 目录下创建 JSP 页面 oneFile.jsp。在该页面中使用表单上传单个文件，具体代码如下：

```jsp
<%@page language="java" contentType="text/html; charset=UTF-8" pageEncoding="UTF-8"%>
<%
String path = request.getContextPath();
String basePath = request.getScheme()+"://"+request.getServerName()+":"+request.getServerPort()+path+"/";
%>
<!DOCTYPE html>
<html>
<head>
<base href="<%=basePath%>">
<meta charset="UTF-8">
<title>Insert title here</title>
<link rel="stylesheet" href="static/css/bootstrap.min.css" />
</head>
<body>
<div class="panel panel-primary">
    <div class="panel-heading">
        <h3 class="panel-title">文件上传示例</h3>
    </div>
</div>
<div class="container">
    <div class="row">
        <div class="col-md-6 col-sm-6">
        <form class="form-horizontal" action="onefile " method="post" enctype="multipart/form-data">
                <div class="form-group">
                    <div class="input-group col-md-6">
                        <span class="input-group-addon">
                            <i class="glyphicon glyphicon-pencil"></i>
                        </span>
                        <input class="form-control" type="text"
                         name="description" placeholder="文件描述"/>
                    </div>
                </div>
                <div class="form-group">
                    <div class="input-group col-md-6">
                        <span class="input-group-addon">
                            <i class="glyphicon glyphicon-search"></i>
```

```html
                </span>
                <input class="form-control" type="file"
                 name="myfile" placeholder="请选择文件"/>
            </div>
        </div>
        <div class="form-group">
            <div class="col-md-6">
                <div class="btn-group btn-group-justified">
                    <div class="btn-group">
                        <button type="submit" class="btn btn-
                         success">
                            <span class="glyphicon glyphicon-share">
                            </span>
                             上传文件
                        </button>
                    </div>
                </div>
            </div>
        </div>
    </form>
  </div>
 </div>
</div>
</body>
</html>
```

4. 创建 POJO 类

在 src/main/java 目录下创建一个名为 pojo 的包，在该包中创建 POJO 类 FileDomain。在该 POJO 类中声明一个 MultipartFile 类型的属性，封装被上传的文件信息，属性名与文件选择页面 oneFile.jsp 中的 file 类型的表单参数名 myfile 相同。具体代码如下：

```java
package pojo;
import org.springframework.web.multipart.MultipartFile;
public class FileDomain {
    private String description;
    private MultipartFile myfile;
    //省略 setter 和 getter 方法
}
```

5. 创建控制器类

在 src/main/java 目录下创建一个名为 controller 的包，并在该包中创建

FileUploadController 控制器类。具体代码如下：

```java
package controller;
import java.io.File;
import javax.servlet.http.HttpServletRequest;
import org.apache.commons.logging.Log;
import org.apache.commons.logging.LogFactory;
import org.springframework.stereotype.Controller;
import org.springframework.web.bind.annotation.ModelAttribute;
import org.springframework.web.bind.annotation.RequestMapping;
import pojo.FileDomain;
@Controller
public class FileUploadController {
    private static final Log logger = LogFactory.getLog(FileUploadController.class);
    /**
     * 单文件上传
     */
    @RequestMapping("/onefile")
    public String oneFileUpload(@ModelAttribute FileDomain fileDomain, HttpServletRequest request){
        /*文件上传到服务器的位置"/uploadfiles",该位置是指
          workspace\.metadata\.plugins\org.eclipse.wst.server.core\tmp0\wtpwebapps,发布后使用*/
        String realpath = request.getServletContext().getRealPath("uploadfiles");
        String fileName = fileDomain.getMyfile().getOriginalFilename();
        File targetFile = new File(realpath, fileName);
        if(!targetFile.exists()){
            targetFile.mkdirs();
        }
        //上传
        try {
            fileDomain.getMyfile().transferTo(targetFile);
            logger.info("成功");
        } catch (Exception e) {
            e.printStackTrace();
        }
        return "showOne";
    }
}
```

6. 创建成功显示页面

在 src/main/webapp/WEB-INF 目录下创建一个名为 jsp 的文件夹，并在该文件夹

中创建单文件上传成功显示页面 showOne.jsp。具体代码如下：

```
<%@page language="java" contentType="text/html; charset=UTF-8" pageEncoding="UTF-8"%>
<!DOCTYPE html PUBLIC "-//W3C//DTD HTML 4.01 Transitional//EN" "http://www.w3.org/TR/html4/loose.dtd">
<html>
<head>
<meta http-equiv="Content-Type" content="text/html; charset=UTF-8">
<title>Insert title here</title>
</head>
<body>
    ${fileDomain.description }<br>
    <!--fileDomain.getMyfile().getOriginalFilename() -->
    ${fileDomain.myfile.originalFilename }
</body>
</html>
```

7. 测试文件上传

发布 ch10_1 应用到 Tomcat 服务器，在浏览器中通过地址 http://localhost:8080/ch10_1/oneFile.jsp 运行文件选择页面，运行效果如图 10.1 所示。

图 10.1　单文件选择页面

在图 10.1 中选择文件，并输入文件描述，然后单击"提交"按钮上传文件，成功显示图 10.2 所示的效果。

图 10.2　单文件成功上传

10.1.5　多文件上传

本小节继续通过 ch10_1 应用案例讲解 Spring MVC 框架如何实现多文件上传。

【例 10-2】　Spring MVC 框架实现多文件上传。

具体实现步骤如下。

1. 创建多文件选择页面

在 src/main/webapp 目录下创建 JSP 页面 multiFiles.jsp。在该页面中使用表单上传多个文件,具体代码如下:

```
<%@page language="java" contentType="text/html; charset=UTF-8" pageEncoding="UTF-8"%>
<%
String path = request.getContextPath();
String basePath = request.getScheme()+"://"+request.getServerName()+":"+request.getServerPort()+path+"/";
%>
<!DOCTYPE html>
<html>
<head>
<base href="<%=basePath%>">
<meta charset="UTF-8">
<title>Insert title here</title>
<link rel="stylesheet" href="static/css/bootstrap.min.css" />
</head>
<body>
<div class="panel panel-primary">
     <div class="panel-heading">
         <h3 class="panel-title">文件上传示例</h3>
     </div>
 </div>
 <div class="container">
     <div class="row">
         <div class="col-md-6 col-sm-6">
<form class="form-horizontal" action="multifile" method="post" enctype="multipart/form-data">
             <div class="form-group">
                 <div class="input-group col-md-6">
                     <span class="input-group-addon">
                         <i class="glyphicon glyphicon-pencil"></i>
                     </span>
                     <input class="form-control" type="text"
                      name="description" placeholder="文件描述 1"/>
                 </div>
             </div>
             <div class="form-group">
                 <div class="input-group col-md-6">
                     <span class="input-group-addon">
```

```html
                    <i class="glyphicon glyphicon-search"></i>
                </span>
                <input class="form-control" type="file"
                 name="myfile" placeholder="请选择文件1"/>
            </div>
        </div>
        <div class="form-group">
            <div class="input-group col-md-6">
                <span class="input-group-addon">
                    <i class="glyphicon glyphicon-pencil"></i>
                </span>
                <input class="form-control" type="text"
                 name="description" placeholder="文件描述2"/>
            </div>
        </div>
        <div class="form-group">
            <div class="input-group col-md-6">
                <span class="input-group-addon">
                    <i class="glyphicon glyphicon-search"></i>
                </span>
                <input class="form-control" type="file"
                 name="myfile" placeholder="请选择文件2"/>
            </div>
        </div>
        <div class="form-group">
            <div class="col-md-6">
                <div class="btn-group btn-group-justified">
                    <div class="btn-group">
                        <button type="submit" class="btn btn-
                         success">
                            <span class="glyphicon glyphicon-share">
                             </span>
                             上传文件
                        </button>
                    </div>
                </div>
            </div>
        </div>
            </form>
        </div>
      </div>
    </div>
</body>
</html>
```

2. 创建POJO类

在pojo包中创建POJO类MultiFileDomain封装多个文件信息，MultiFileDomain类的具体代码如下：

```java
package pojo;
import java.util.List;
import org.springframework.web.multipart.MultipartFile;
public class MultiFileDomain {
    private List<String>description;
    private List<MultipartFile>myfile;
    //省略setter和getter方法
}
```

3. 添加多文件上传处理方法

在控制器类FileUploadController中添加多文件上传的处理方法multiFileUpload()，具体代码如下：

```java
/**
 * 多文件上传
 */
@RequestMapping("/multifile")
public String multiFileUpload(@ModelAttribute MultiFileDomain multiFileDomain,
HttpServletRequest request){
    String realpath = request.getServletContext().getRealPath("uploadfiles");
    File targetDir = new File(realpath);
    if(!targetDir.exists()){
        targetDir.mkdirs();
    }
    List<MultipartFile>files = multiFileDomain.getMyfile();
    for (int i = 0; i < files.size(); i++) {
        MultipartFile file = files.get(i);
        String fileName = file.getOriginalFilename();
        File targetFile = new File(realpath,fileName);
        //上传
        try {
            file.transferTo(targetFile);
        } catch (Exception e) {
            e.printStackTrace();
        }
    }
    logger.info("成功");
    return "showMulti";
```

}

4. 创建成功显示页面

在 jsp 文件夹中创建多文件上传成功显示页面 showMulti.jsp。具体代码如下：

```jsp
<%@page language="java" contentType="text/html; charset=UTF-8" pageEncoding="UTF-8"%>
<%@taglib uri="http://java.sun.com/jsp/jstl/core" prefix="c" %>
<%
String path =request.getContextPath();
String basePath =   request.getScheme()+"://"+request.getServerName()+":"+request.getServerPort()+path+"/";
%>
<!DOCTYPE html>
<html>
<head>
<base href="<%=basePath%>">
<meta charset="UTF-8">
<title>Insert title here</title>
<link rel="stylesheet" href="static/css/bootstrap.min.css" />
</head>
<body>
    <div class="container">
        <div class="panel panel-primary">
            <div class="panel-heading">
                <h3 class="panel-title">文件列表</h3>
            </div>
            <div class="panel-body">
                <div class="table table-responsive">
                    <table class="table table-bordered table-hover">
                        <tbody class="text-center">
                            <tr>
                                <th>详情</th>
                                <th>文件名</th>
                            </tr>
                            <!--同时取两个数组的元素 -->
            <c:forEach items="${multiFileDomain.description}" var="description"
                varStatus="loop">
                                <tr>
                                    <td>${description}</td>
                        <td>${multiFileDomain.myfile[loop.count-1].
                        originalFilename}</td>
                                </tr>
                            </c:forEach>
```

```
                    </tbody>
                </table>
            </div>
        </div>
    </div>
</body>
</html>
```

5. 测试文件上传

发布 ch10_1 应用到 Tomcat 服务器，在浏览器中通过地址 http://localhost:8080/ch10_1/multiFiles.jsp 运行多文件选择页面，运行效果如图 10.3 所示。

图 10.3　多文件选择页面

在图 10.3 中选择文件，并输入文件描述，然后单击"上传文件"按钮上传多个文件，成功显示图 10.4 所示的效果。

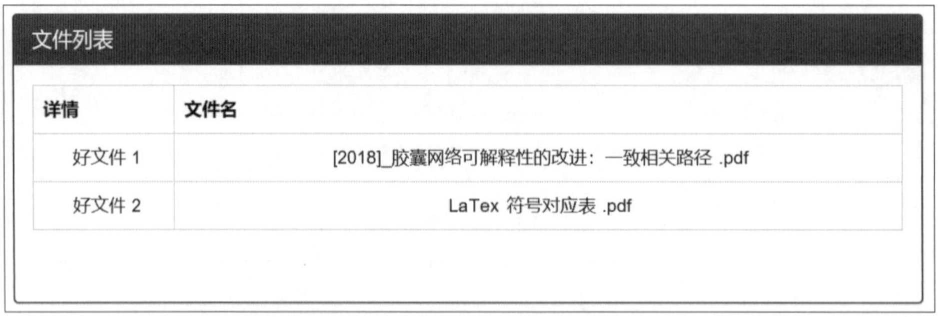

图 10.4　多文件成功上传结果

10.1.6　实践环节

对 10.1.5 小节多文件上传实例的文件类型做限定，例如：只允许上传 .bmp、.gif、

.jpg、.ico 等类型的文件。

10.2 文件下载

10.2.1 文件下载的实现方法

实现文件下载经常有两种方法：一种是通过超链接实现下载；另一种是利用程序编码实现下载。超链接实现下载固然简单，但暴露了下载文件的真实位置，并且只能下载存放在 Web 应用程序所在的目录下的文件。利用程序编码实现下载既可以增加安全访问控制，还可以从任意位置提供下载的数据，既可以将文件存放到 Web 应用程序以外的目录中，也可以将文件保存到数据库中。

利用程序实现下载需要设置两个报头：

（1）Web 服务器需要告诉浏览器其所输出内容的类型不是普通文本文件或 HTML 文件，而是一个要保存到本地的下载文件。设置 Content-Type 的值为：application/x-msdownload。

（2）Web 服务器希望浏览器不直接处理相应的实体内容，而是由用户选择将相应的实体内容保存到一个文件中，这需要设置 Content-Disposition 报头。该报头指定了接收程序处理数据内容的方式，在 HTTP 应用中只有 attachment 是标准方式，attachment 表示要求用户干预。在 attachment 后面还可以指定 filename 参数，该参数是服务器建议浏览器将实体内容保存到文件中的文件名称。

设置报头的示例如下：

```
response.setHeader("Content-Type", "application/x-msdownload");
response.setHeader("Content-Disposition", "attachment; filename=" +filename);
```

10.2.2 文件下载

下面继续通过应用 ch10_1 讲述利用程序实现下载的过程。要求从 10.1 节上传文件的目录 workspace\.metadata\.plugins\org.eclipse.wst.server.core\tmp0\wtpwebapps\ch10_1\uploadfiles 中下载文件。

【例 10-3】 Spring MVC 框架实现文件下载。

具体实现步骤如下。

1. 编写控制器类

首先，在包 controller 中编写控制器类 FileDownController，该类中有 3 个方法：show、down 和 toUTF8String。show 方法获取被下载的文件名称；down 方法执行下载功能；toUTF8String 方法是下载保存时中文文件名字符编码转换方法。FileDownController 类的代码如下：

```java
package controller;
import java.io.File;
import java.io.FileInputStream;
import java.io.UnsupportedEncodingException;
import java.util.ArrayList;
import javax.servlet.ServletOutputStream;
import javax.servlet.http.HttpServletRequest;
import javax.servlet.http.HttpServletResponse;
import org.apache.commons.logging.Log;
import org.apache.commons.logging.LogFactory;
import org.springframework.stereotype.Controller;
import org.springframework.ui.Model;
import org.springframework.web.bind.annotation.RequestMapping;
import org.springframework.web.bind.annotation.RequestParam;
@Controller
public class FileDownController {
    private static final Log logger =LogFactory.getLog(FileDownController.class);
    /**
     * 显示要下载的文件
     */
    @RequestMapping("showDownFiles")
    public String show(HttpServletRequest request, Model model){
        //从 workspace\.metadata\.plugins\org.eclipse.wst.server.core\tmp0\wtpwebapps\ch10_1\下载
        String realpath =request.getServletContext().getRealPath("uploadfiles");
        File dir =new File(realpath);
        File files[] =dir.listFiles();
        //获取该目录下的所有文件名
        ArrayList<String>fileName =new ArrayList<String>();
        for (int i =0; i <files.length; i++) {
            fileName.add(files[i].getName());
        }
        model.addAttribute("files", fileName);
        return "showDownFiles";
    }
    /**
     * 执行下载
     */
    @RequestMapping("down")
    public String down(@RequestParam String filename, HttpServletRequest request, HttpServletResponse response){
        String aFilePath =null; //要下载的文件路径
        FileInputStream in =null; //输入流
        ServletOutputStream out =null; //输出流
```

```java
        try {
            //从\tmp0\wtpwebapps\ch10_1\uploadfiles 下载
            aFilePath = request.getServletContext().getRealPath("uploadfiles");
            //设置下载文件使用的报头
            response.setHeader("Content-Type", "application/x-msdownload" );
            response.setHeader("Content-Disposition", "attachment; filename="+
                toUTF8String(filename));
            // 读入文件
            in = new FileInputStream(aFilePath + "\\"+filename);
            //得到响应对象的输出流,用于向客户端输出二进制数据
            out = response.getOutputStream();
            out.flush();
            int aRead = 0;
            byte b[] = new byte[1024];
            while ((aRead = in.read(b)) != -1 & in != null) {
                out.write(b, 0, aRead);
            }
            out.flush();
            in.close();
            out.close();
        } catch (Throwable e) {
            e.printStackTrace();
        }
        logger.info("下载成功");
        return null;
    }
    /**
     * 下载保存时中文文件名字符编码转换方法
     */
    public  String toUTF8String(String str){
        StringBuffer sb = new StringBuffer();
        int len = str.length();
        for(int i = 0; i < len; i++){
            //取出字符中的每个字符
            char c = str.charAt(i);
            //Unicode 码值在 0~255,不作处理
            if(c >= 0 && c <= 255){
                sb.append(c);
            }else{//转换 UTF-8 编码
                byte b[];
                try {
                    b = Character.toString(c).getBytes("UTF-8");
                } catch (UnsupportedEncodingException e) {
                    e.printStackTrace();
```

```java
                b = null;
            }
            //转换为%HH的字符串形式
            for(int j = 0; j < b.length; j ++){
                int k = b[j];
                if(k < 0){
                    k &= 255;
                }
                sb.append("%" + Integer.toHexString(k).toUpperCase());
            }
        }
    }
    return sb.toString();
}
```

2. 创建文件列表页面

在jsp文件夹中创建一个显示被下载文件的JSP页面showDownFiles.jsp,代码如下:

```jsp
<%@page language="java" contentType="text/html; charset=UTF-8" pageEncoding="UTF-8"%>
<%@taglib uri="http://java.sun.com/jsp/jstl/core" prefix="c" %>
<%
String path = request.getContextPath();
String basePath = request.getScheme() + "://" + request.getServerName() + ":" + request.getServerPort()+path+"/";
%>
<!DOCTYPE html>
<html>
<head>
<base href="<%=basePath%>">
<meta charset="UTF-8">
<title>Insert title here</title>
<link rel="stylesheet" href="static/css/bootstrap.min.css" />
</head>
<body>
    <div class="container">
        <div class="panel panel-primary">
            <div class="panel-heading">
                <h3 class="panel-title">文件列表</h3>
            </div>
            <div class="panel-body">
```

```
            <div class="table table-responsive">
                <table class="table table-bordered table-hover">
                    <tbody class="text-center">
                        <tr>
                            <th>被下载的文件名</th>
                            <th>下载操作</th>
                        </tr>
                        <!--遍历 model 中的 files -->
                        <c:forEach items="${files}" var="filename">
                            <tr>
                                <td>${filename}</td>
<td><a href="down?filename=${filename}">DownLoad
                                </a></td>
                            </tr>
                        </c:forEach>
                    </tbody>
                </table>
            </div>
        </div>
    </div>
</body>
</html>
```

3. 测试下载功能

发布 ch10_1 应用到 Tomcat 服务器。然后，在浏览器中通过地址 http://localhost:8080/ch10_1/showDownFiles 测试下载示例，运行效果如图 10.5 所示。

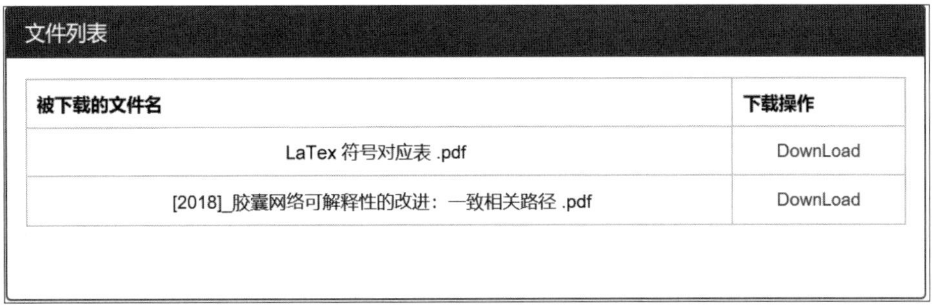

文件列表	
被下载的文件名	下载操作
LaTex 符号对应表 .pdf	DownLoad
[2018]_胶囊网络可解释性的改进：一致相关路径 .pdf	DownLoad

图 10.5　被下载文件列表

单击图 10.5 中的 DownLoad 超链接，下载文件。需要注意的是，使用浏览器演示案例，不能在 STS 中演示下载案例。

10.3 本章小结

本章重点介绍了 Spring MVC 的文件上传，主要包括如何使用 MultipartFile 接口封装文件信息。最后介绍了如何使用 Spring MVC 框架编程进行文件下载。

习 题 10

1. 基于表单的文件上传，应将表单的 enctype 属性值设置为（　　）。
   A. multipart/form-data
   B. application/x-www-form-urlencoded
   C. text/plain
   D. html/text
2. 在 Spring MVC 框架中，如何限定上传文件的大小？
3. 单文件上传与多文件上传有什么区别？

EL 与 JSTL

学习目的与要求

本章主要介绍表达式语言(Expression Language, EL)和 JSP 标准标签库(Java Server Pages Standard Tag Library, JSTL)的基本用法。通过本章的学习,掌握 EL 表达式语法,掌握 EL 隐含对象,了解什么是 JSTL,掌握 JSTL 的核心标签库。

本章主要内容

- EL。
- JSTL。

在 JSP 页面中,可以使用 Java 代码实现页面显示逻辑,但网页中夹杂着 HTML 与 Java 代码,给网页的设计与维护带来困难。可以使用 EL 来访问和处理应用程序的数据,也可以使用 JSTL 来替换网页中实现页面显示逻辑的 Java 代码。这样 JSP 页面就尽量减少了 Java 代码的使用,为以后的维护提供方便。

11.1 表达式语言 EL

EL 是 JSP2.0 规范中增加的,它的基本语法如下:

${表达式}

EL 表达式类似 JSP 表达式<%=表达式%>,EL 语句中的表达式值将被直接送到浏览器显示。通过 page 指令的 isELIgnored 属性来说明是否支持 EL 表达式。isELIgnored 属性值为 false 时,JSP 页面可以使用 EL 表达式;isELIgnored 属性值为 true 时,JSP 页面不能使用 EL 表达式。isELIgnored 的属性值默认为 false。

11.1.1 基本语法

EL 的语法简单,使用方便。它以"${"开始,以"}"结束。

1. "[ ]"与"."运算符

EL 使用"[ ]"和"."运算符来访问数据,主要使用 EL 获取对象的属性,包括获取 JavaBean 的属性值、获取数组中的元素以及获取集合对象中的元素。对于 null 值,直接以空字符串显示,而不是 null,运算时也不会发生错误或空指针异常。所以在使用 EL 访问对象属性时,不需判断对象是否为 null 对象。

(1) 获取 JavaBean 的属性值。

假设在 JSP 页面中有这样一段代码:

```
<%=user.getAge ()%>
```

那么,可以使用 EL 获取 user 对象的属性 age,代码如下:

```
${user.age}
```

或

```
${user["age"]}
```

其中,点运算符前面为 JavaBean 的对象 user,后面为该对象的属性 age,表示利用 user 对象的 getAge()方法取值,并显示在网页上。

(2) 获取数组中的元素。

假设在 Controller 控制器中有这样一段话:

```
String dogs[] ={"lili","huahua","guoguo"};
request.setAttribute("array", dogs);
```

那么,在对应视图 JSP 中可以使用 EL 取出数组中的元素(也可以使用 11.2 节的 JSTL 遍历数组),代码如下:

```
${array[0]}
```

${array[1]}

${array[2]}

(3) 获取集合对象中的元素。

假设在 Controller 控制器中有这样一段代码：

```
ArrayList<UserBean>users =new ArrayList<UserBean>();
UserBean ub1 =new UserBean("zhang",20);
UserBean ub2 =new UserBean("zhao",50);
users.add(ub1);
users.add(ub2);
request.setAttribute("array", users);
```

其中，UserBean 有两个属性：name 和 age，在对应视图 JSP 页面中可以使用 EL 取出 UserBean 中的属性(也可以使用 11.2 节的 JSTL 遍历数组)，代码如下：

${array[0].name}　${array[0].age}

${array[1].name}　${array[1].age}

2. 算术运算符

EL 表达式中有 5 个算术运算符，如表 11.1 所示。

表 11.1　EL 的算术运算符

算术运算符	说　　明	示　　例	结　　果
＋	加	${13＋2}	15
－	减	${13－2}	11
*	乘	${13*2}	26
/(或 div)	除	${13/2}或 ${13 div 2}	6.5
％(或 mod)	取模(求余)	${13％2}或 ${13 mod 2}	1

3. 关系运算符

EL 表达式中有 6 个关系运算符，如表 11.2 所示。

表 11.2　EL 的关系运算符

关系运算符	说　　明	示　　例	结　　果
==(或 eq)	等于	${13 == 2}或 ${13 eq 2}	false
!=(或 ne)	不等于	${13 != 2}或 ${13 ne 2}	true
<(或 lt)	小于	${13 < 2}或 ${13 lt 2}	false
>(或 gt)	大于	${13 > 2}或 ${13 gt 2}	true

续表

关系运算符	说明	示例	结果
<=(或 le)	小于等于	${13 <= 2}$或${13 le 2}$	false
>=(或 ge)	大于等于	${13 >= 2}$或${13 ge 2}$	true

4. 逻辑运算符

EL 表达式中有 3 个逻辑运算符，如表 11.3 所示。

表 11.3 EL 的逻辑运算符

逻辑运算符	说明	示例	结果
&&(或 and)	逻辑与	如果 A 为 true,B 为 false,则 A && B(或 A and B)	false
||(或 or)	逻辑或	如果 A 为 true,B 为 false,则 A || B(或 A or B)	true
！(或 not)	逻辑非	如果 A 为 true,则！A(或 not A)	false

5. empty 运算符

empty 运算符用于检测一个值是否为 null，例如，变量 A 不存在，则 ${empty A}$ 返回的结果为 true。

6. 条件运算符

EL 中的条件运算符是"？:"，例如，${A ? B:C}$，如果 A 为 true，计算 B 并返回其结果，如果 A 为 false，计算 C 并返回其结果。

11.1.2 EL 隐含对象

EL 的隐含对象共有 11 个，本书只介绍几个常用的 EL 隐含对象：
pageScope、requestScope、sessionScope、applicationScope、param 以及 paramValues。

1. 与作用范围相关的隐含对象

与作用范围有关的 EL 隐含对象有 pageScope、requestScope、sessionScope 和 applicationScope，分别可以获取 JSP 隐含对象 pageContext、request、session 和 application 中的数据。如果在 EL 中没有使用隐含对象指定作用范围，则会依次从 page、request、session、application 范围查找，找到就直接返回，不再继续找下去，如果所有范围都没有找到，就返回空字符串。获取数据的格式如下：

${EL 隐含对象.关键字对象.属性}

或

${EL 隐含对象.关键字对象}

例如：

```
<jsp:useBean id="user" class="bean.UserBean" scope="page"/><!--bean 标签-->
<jsp:setProperty name="user" property="name" value="EL 隐含对象" />
name: ${pageScope.user.name}
```

再如，Controller 控制器中有这样一段代码：

```
ArrayList<UserBean>users =new ArrayList<UserBean>();
UserBean ub1 =new UserBean("zhang",20);
UserBean ub2 =new UserBean("zhao",50);
users.add(ub1);
users.add(ub2);
request.setAttribute("array", users);
```

其中，UserBean 有两个属性：name 和 age，那么在对应视图 JSP 中，request 有效的范围内可以使用 EL 取出 UserBean 的属性（也可以使用 11.2 节的 JSTL 遍历数组），代码如下：

```
${requestScope.array[0].name}  ${requestScope.array[0].age}
${requestScope.array[1].name}  ${requestScope.array[1].age}
```

2. 与请求参数相关的隐含对象

与请求参数相关的 EL 隐含对象有 param 和 paramValues。获取数据的格式如下：

`${EL 隐含对象.参数名}`

比如，input.jsp 的代码如下：

```
<form method ="post" action ="param.jsp">
    <p>姓名：<input type="text" name="username" size="15" /></p>
    <p>兴趣：
    <input type="checkbox" name="habit" value="看书"/>看书
    <input type="checkbox" name="habit" value="玩游戏"/>玩游戏
    <input type="checkbox" name="habit" value="旅游"/>旅游
    <p>
    <input type="submit" value="提交"/>
</form>
```

那么，在 param.jsp 页面中可以使用 EL 获取参数值，代码如下：

```
<%request.setCharacterEncoding("GBK");%>
<body>
<h2>EL 隐含对象 param、paramValues</h2>
姓名：${param.username}</br>
兴趣：
${paramValues.habit[0]}
```

```
${paramValues.habit[1]}
${paramValues.habit[2]}
```

【例 11-1】 编写一个 Controller,在该控制器类处理方法中使用 request 对象和 Model 对象存储数据,然后从处理方法转发到 show.jsp 页面,show.jsp 页面中显示 request 和 Model 对象的数据。

具体实现步骤如下。

1. 创建 Maven 项目并添加依赖的 JAR 包

创建一个名为 ch11_1 的 Maven 项目,并通过 pom.xml 文件添加项目的相关依赖 spring-webmvc。pom.xml 文件的内容如下:

```xml
<project xmlns="http://maven.apache.org/POM/4.0.0"
xmlns:xsi="http://www.w3.org/2001/XMLSchema-instance"
xsi:schemaLocation="http://maven.apache.org/POM/4.0.0 https://maven.apache.org/xsd/maven-4.0.0.xsd">
  <modelVersion>4.0.0</modelVersion>
  <groupId>com.cn.chenheng</groupId>
  <artifactId>ch11_1</artifactId>
  <version>0.0.1-SNAPSHOT</version>
  <packaging>war</packaging>
  <properties>
    <!--spring 版本号 -->
    <spring.version>5.2.3.RELEASE</spring.version>
  </properties>
  <dependencies>
    <dependency>
      <groupId>org.springframework</groupId>
      <artifactId>spring-webmvc</artifactId>
      <version>${spring.version}</version>
    </dependency>
  </dependencies>
</project>
```

2. 创建 Web 和 Spring MVC 的 Java 配置类

在 ch11_1 应用的 src/main/java 目录下创建一个名为 config 的包,并在该包中创建 Web 的配置类 WebConfig 和 Spring MVC 的配置类 SpringMVCConfig。

WebConfig 的代码如下:

```java
package config;
import javax.servlet.ServletContext;
import javax.servlet.ServletException;
import javax.servlet.ServletRegistration.Dynamic;
```

```java
import org.springframework.web.WebApplicationInitializer;
import org.springframework.web.context.support.
AnnotationConfigWebApplicationContext;
import org.springframework.web.servlet.DispatcherServlet;
public class WebConfig implements WebApplicationInitializer{
    @Override
    public void onStartup(ServletContext arg0) throws ServletException {
        AnnotationConfigWebApplicationContext ctx
        =new AnnotationConfigWebApplicationContext();
         ctx.register(SpringMVCConfig.class);//注册 Spring MVC 的 Java 配置类
                                              SpringMVCConfig
        ctx.setServletContext(arg0);//和当前 ServletContext 关联
        /**
         * 注册 Spring MVC 的 DispatcherServlet
         */
        Dynamic servlet =arg0.addServlet("dispatcher", new DispatcherServlet
                                    (ctx));
        servlet.addMapping("/");
        servlet.setLoadOnStartup(1);
    }
}
```

SpringMVCConfig 的代码如下：

```java
package config;
import org.springframework.context.annotation.Bean;
import org.springframework.context.annotation.ComponentScan;
import org.springframework.context.annotation.Configuration;
import org.springframework.web.servlet.config.annotation.EnableWebMvc;
import org.springframework.web.servlet.config.annotation.ResourceHandlerRegistry;
import org.springframework.web.servlet.config.annotation.WebMvcConfigurer;
import org.springframework.web.servlet.view.InternalResourceViewResolver;
@Configuration
@EnableWebMvc // 开启 Spring MVC 的支持
@ComponentScan(basePackages ={ "controller"}) // 扫描基本包
public class SpringMVCConfig implements WebMvcConfigurer {
    /**
     * 配置视图解析器
     */
    @Bean
    public InternalResourceViewResolver getViewResolver() {
        InternalResourceViewResolver viewResolver =new
        InternalResourceViewResolver();
        viewResolver.setPrefix("/WEB-INF/jsp/");
        viewResolver.setSuffix(".jsp");
```

```
        return viewResolver;
    }
    /**
     * 配置静态资源(不需要DispatcherServlet转发的请求)
     */
    @Override
    public void addResourceHandlers(ResourceHandlerRegistry registry) {
        registry.addResourceHandler("/static/**").addResourceLocations
            ("/static/");
    }
}
```

3. 创建控制器类

在 ch11_1 应用的 src/main/java 目录下创建一个名为 controller 的包,并在该包中创建一个名为 InputController 的控制器类。代码如下:

```
package controller;
import javax.servlet.http.HttpServletRequest;
import org.springframework.stereotype.Controller;
import org.springframework.ui.Model;
import org.springframework.web.bind.annotation.RequestMapping;
@Controller
public class InputController {
    @RequestMapping("/input")
    public String input(HttpServletRequest request, Model model){
        String names[] ={ "zhao", "qian", "sun", "li" };
        request.setAttribute("name", names);
        String address[] ={ "beijing", "shanghai", "shenzhen"};
        model.addAttribute("address", address);
        return "show";
    }
}
```

4. 创建 show.jsp 页面

在 src/main/webapp/WEB-INF 目录下创建 jsp 文件夹,在该文件夹中创建 show.jsp 页面。代码如下:

```
<%@page language="java" contentType="text/html; charset=UTF-8" pageEncoding="UTF-8"%>
<%
String path = request.getContextPath();
String basePath = request.getScheme()+"://"+request.getServerName()+":"+request.getServerPort()+path+"/";
```

```
%>
<!DOCTYPE html>
<html>
<head>
<base href="<%=basePath%>">
<meta charset="UTF-8">
<title>Insert title here</title>
</head>
<body>
    从Controller转发过来的request内置对象的数据如下:<br>
    ${requestScope.name[0]}<br>
    ${requestScope.name[1]}<br>
    ${requestScope.name[2]}<br>
    ${requestScope.name[3]}<br>
    <hr>
    从Controller转发过来的Model对象的数据如下:<br>
    ${address[0]}<br>
    ${address[1]}<br>
    ${address[2]}<br>
</body>
</html>
```

5. 测试运行

首先，发布应用到 Tomcat 服务器。然后，在浏览器的地址栏中输入 http://localhost:8080/ch11_1/input，程序运行结果如图 11.1 所示。

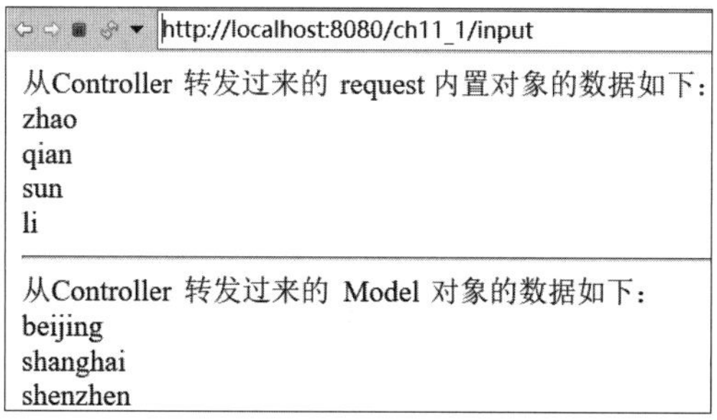

图 11.1　使用 EL 获取数据

11.1.3　实践环节

将【例 11-1】的 Controller 控制器类的 input 方法修改如下：

```
@RequestMapping("input")
public String input(HttpServletRequest request){
    Map<String, String>names =new HashMap<String, String>();
```

```
        names.put("first", "zhao");
        names.put("second", "qian");
        names.put("third", "sun");
        names.put("forth", "li");
        request.setAttribute("name", names);
        return "show";
}
```

请修改【例 11-1】的 show.jsp，显示 map 中的数据。

11.2　JSP 标准标签库 JSTL

　　JSTL 规范由 Sun 公司制定，Apache 的 Jakarta 小组负责实现。JSTL 标准标签库由 5 个不同功能的标签库组成，包括 Core、I18N、XML、SQL 以及 Functions，本节只是简要介绍了 JSTL 的 Core 和 Functions 标签库中几个常用的标签。

11.2.1　配置 JSTL

　　JSTL 现在已经是 Java EE5 的一个组成部分了，如果采用支持 Java EE5 或以上版本的集成开发环境开发 Web 应用程序时，就不需要再配置 JSTL 了。但本书采用的是 STS 平台，因此需要配置 JSTL。配置 JSTL 的步骤如下。

1. 复制 JSTL 的标准实现

　　如果不使用 Maven 添加 JSTL 的依赖，那么可以在 Tomcat 的\webapps\examples\WEB-INF\ lib 目录下找到 "taglibs-standard-impl-1.2.5.jar" 和 "taglibs-standard-spec-1.2.5.jar" 文件，然后复制到 Web 工程的 WEB-INF\lib 目录下即可。

　　如果使用 Maven 添加 JSTL 的依赖，那么可以通过应用的 pom.xml 文件添加，代码如下：

```
<dependency>
    <groupId>javax.servlet</groupId>
    <artifactId>jstl</artifactId>
    <version>1.2</version>
</dependency>
```

2. 使用 taglib 标记定义前缀与 uri 引用

　　如果使用 Core 标签库，需要在 JSP 页面中使用 taglib 标记定义前缀与 uri 引用，代码如下：

```
<%@taglib prefix="c" uri="http://java.sun.com/jsp/jstl/core"%>
```

　　如果使用 Functions 标签库，需要在 JSP 页面中使用 taglib 标记定义前缀与 uri 引

用,代码如下:

```
<%@taglib prefix="fn" uri="http://java.sun.com/jsp/jstl/functions"%>
```

11.2.2 核心标签库之通用标签

1. <c:out> 标签

<c:out>用来显示数据的内容,与<%= 表达式 %>或${表达式}类似。格式如下:

```
<c:out value="输出的内容" [default="defaultValue"]/>
```

或

```
<c:out value="输出的内容">
    defaultValue
</c:out>
```

其中,value 值可以是一个 EL 表达式,也可以是一个字符串;default 可有可无,当 value 值不存在时,就输出 defaultValue。例如:

```
<c:out value="${param.data}" default="没有数据" />
<br>
<c:out value="${param.nothing}" />
<br>
<c:out value="这是一个字符串" />
```

以上代码输出结果如图 11.2 所示。

图 11.2 <c:out>标签

2. <c:set> 标签

(1) 设置作用域变量。

可以使用<c:set>在 page、request、session、application 等范围内设置一个变量。格式如下:

```
<c:set value="value" var="varName" [scope="page|request|session|application"]/>
```

将 value 值赋值给变量 varName。例如:

```
<c:set value="zhao" var="userName" scope="session"/>
```

相当于

```
<%session.setAttribute("userName","zhao"); %>
```

(2) 设置 JavaBean 的属性。

使用<c:set>设置 JavaBean 的属性时,必须使用 target 属性进行设置。格式如下:

```
<c:set value="value" target="target"  property="propertyName"/>
```

将 value 赋值给 target 对象(JaveBean 对象)的 propertyName 属性。如果 target 为 null,或没有 set 方法,则抛出异常。

3. < c:remove> 标签

如果要删除某个变量,可以使用<c:remove>标签。例如:

```
<c:remove var="userName" scope="session"/>
```

相当于

```
<%session.removeAttribute("userName") %>
```

11.2.3 核心标签库之流程控制标签

1. < c:if> 标签

<c:if>标签实现 if 语句的作用,具体语法格式如下:

```
<c:if test="条件表达式">
    主体内容
</c:if>
```

其中,条件表达式可以是 EL 表达式,也可以是 JSP 表达式。如果表达式的值为 true,则执行<c:if>的主体内容,但是没有相对应的<c:else>标签。如果想在条件成立时执行一块内容,不成立时执行另一块内容,则可以使用<c:choose>、<c:when>及<c:otherwise>标签。

2. < c:choose> 、< c:when> 及< c:otherwise> 标签

<c:choose>、<c:when>及<c:otherwise>标签实现 if/elseif/else 语句的作用。具体语法格式如下:

```
<c:choose>
    <c:when test="条件表达式 1">
        主体内容 1
    </c:when>
    <c:when test="条件表达式 2">
        主体内容 2
    </c:when>
```

```
    <c:otherwise>
        表达式都不正确时,执行的主体内容
    </c:otherwise>
</c:choose>
```

【例 11-2】 编写一个 JSP 页面 ifelse.jsp,在该页面中使用＜c:set＞标签,把两个字符串设置为 request 范围内的变量。使用＜c:if＞标签求出这两个字符串的最大值(按字典顺序比较大小),使用＜c:choose＞、＜c:when＞及＜c:otherwise＞标签求出这两个字符串的最小值。

例 11-2 页面文件 ifelse.jsp 的代码如下：

```
<%@page language="java" contentType="text/html; charset=UTF-8" pageEncoding=
"UTF-8"%>
<%@taglib uri="http://java.sun.com/jsp/jstl/core" prefix="c"%>
<%
    String path =request.getContextPath();
    String basePath =request.getScheme() +"://" +request.getServerName() +":"
    +request.getServerPort() +path +"/";
%>
<!DOCTYPE html>
<html>
<head>
<base href="<%=basePath%>">
<meta charset="UTF-8">
<title>Insert title here</title>
</head>
<body>
    <c:set value="if" var="firstNumber" scope="request" />
    <c:set value="else" var="secondNumber" scope="request" />
    <c:if test="${firstNumber>secondNumber}">
        最大值为${firstNumber}
    </c:if>
    <c:if test="${firstNumber<secondNumber}">
        最大值为${secondNumber}
    </c:if>
    <c:choose>
        <c:when test="${firstNumber<secondNumber}">
            最小值为${firstNumber}
        </c:when>
        <c:otherwise>
            最小值为${secondNumber}
        </c:otherwise>
    </c:choose>
</body>
```

</html>

<c:when>及<c:otherwise>必须放在<c:choose>之中。当<c:when>的 test 结果为 true 时,会输出<c:when>的主体内容,而不理会<c:otherwise>的内容。<c:choose>中可有多个<c:when>,程序会从上到下进行条件判断,如果某个<c:when>的 test 结果为 true,就输出其主体内容,之后的<c:when>就不再执行。如果所有的<c:when>的 test 结果都为 false 时,则会输出<c:otherwise>的内容。<c:if>与<c:choose>也可以嵌套使用,例如:

```
<c:set value="fda" var="firstNumber" scope="request"/>
<c:set value="else" var="secondNumber" scope="request"/>
<c:set value="ddd" var="threeNumber" scope="request"/>
<c:if test="${firstNumber>secondNumber}">
    <c:choose>
        <c:when test="${firstNumber<threeNumber}">
            最大值为${threeNumber}
        </c:when>
        <c:otherwise>
            最大值为${firstNumber}
        </c:otherwise>
    </c:choose>
</c:if>
<c:if test="${secondNumber>firstNumber}">
    <c:choose>
        <c:when test="${secondNumber<threeNumber}">
            最大值为${threeNumber}
        </c:when>
        <c:otherwise>
            最大值为${secondNumber}
        </c:otherwise>
    </c:choose>
</c:if>
```

11.2.4 核心标签库之迭代标签

1. < c:forEach> 标签

<c:forEach>标签可以实现程序中的 for 循环。语法格式如下:

```
<c:forEach var="变量名" items="数组或 Collection 对象">
    循环体
</c:forEach>
```

其中,items 属性可以是数组或 Collection 对象,每次循环读取对象中的一个元素,并赋值给 var 属性指定的变量,之后就可以在循环体使用 var 指定的变量获取对象的元素。例如，Controller 控制器中有这样一段代码：

```
ArrayList<UserBean>users =new ArrayList<UserBean>();
UserBean ub1 =new UserBean("zhao",20);
UserBean ub2 =new UserBean("qian",40);
UserBean ub3 =new UserBean("sun",60);
UserBean ub4 =new UserBean("li",80);
users.add(ub1);
users.add(ub2);
users.add(ub3);
users.add(ub4);
request.setAttribute("usersKey", users);
```

那么，在对应JSP页面中可以使用＜c:forEach＞循环遍历出数组中的元素。代码如下：

```
<table>
  <tr>
    <th>姓名</th>
    <th>年龄</th>
  </tr>
<c:forEach var="user" items="${requestScope.usersKey}">
  <tr>
    <td>${user.name}</td>
    <td>${user.age}</td>
  </tr>
</c:forEach>
</table>
```

有些情况下，需要为＜c:forEach＞标签指定 begin、end、step 和 varStatus 属性。begin 为迭代时的开始位置，默认值为 0；end 为迭代时的结束位置，默认值是最后一个元素；step 为迭代步长，默认值为 1；varStatus 代表迭代变量的状态，包括 count（迭代的次数）、index（当前迭代的索引，第一个索引为 0）、first（是否是第一个迭代对象）和 last（是否是最后一个迭代对象）。例如：

```
<%@page language="java" contentType="text/html; charset=UTF-8" pageEncoding=
"UTF-8"%>
<%@taglib prefix="c" uri="http://java.sun.com/jsp/jstl/core"%>
<%
    String path =request.getContextPath();
    String basePath =request.getScheme() +"://" +request.getServerName() +":"
    +request.getServerPort()+path +"/";
%>
<!DOCTYPE html>
<html>
<head>
<base href="<%=basePath%>">
```

```html
<meta charset="UTF-8">
<title>Insert title here</title>
<link rel="stylesheet" href="static/css/bootstrap.min.css" />
<link rel="stylesheet" href="static/css/bootstrap-theme.min.css" />
</head>
<body>
    <div class="container">
        <div class="panel panel-primary">
            <div class="panel-body">
                <div class="table table-responsive">
                    <table class="table table-bordered table-hover">
                        <tbody class="text-center">
                            <tr>
                                <th>Value</th>
                                <th>Square</th>
                                <th>Index</th>
                            </tr>
                            <c:forEach var="x" varStatus="status" begin="0" end="10" step="2">
                                <tr>
                                    <td>${x}</td>
                                    <td>${x * x}</td>
                                    <td>${status.index}</td>
                                </tr>
                            </c:forEach>
                        </tbody>
                    </table>
                </div>
            </div>
        </div>
    </div>
</body>
</html>
```

上述程序运行结果如图11.3所示。

2. ＜c:forTokens＞标签

＜c:forTokens＞用于迭代字符串中由分隔符分隔的各成员，它通过java.util.StringTokenizer实例来完成字符串的分隔，属性items和delims作为构造StringTokenizer实例的参数。语法格式如下：

```
<c:forTokens var="变量名" items="要迭代的String对象" delims="指定分隔字符串的分隔符">
    循环体
</c:forTokens>
```

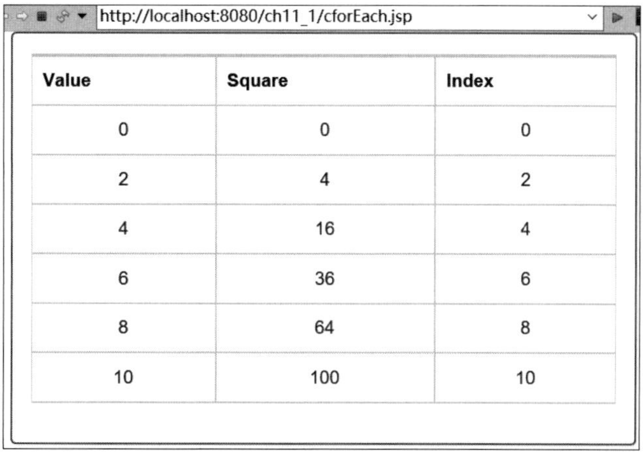

Value	Square	Index
0	0	0
2	4	2
4	16	4
6	36	6
8	64	8
10	100	10

图 11.3　<c:forEach>标签

例如：

```
<c:forTokens items="chenheng1:chenheng2:chenheng3" delims=":" var="name">
      ${name}<br>
</c:forTokens>
```

上述程序运行结果如图 11.4 所示。

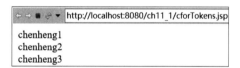

图 11.4　<c:forTokens>标签

<c:forTokens>标签与<c:forEach>标签一样，也有 begin、end、step 和 varStatus 属性，并且用法相同，不再赘述。

11.2.5　函数标签库

在 JSP 页面中调用 JSTL 中的函数时，需要使用 EL 表达式，调用语法格式如下：

```
${fn:函数名(参数 1,参数 2,…)}
```

下面介绍几个常用的函数。

1. contains 函数

该函数功能是判断一个字符串中是否包含指定的子字符串。如果包含，则返回 true，否则返回 false。其定义如下：

```
contains(string, substring)
```

该函数调用示例代码如下：

`${fn:contains("I am studying", "am") }`

上述 EL 表达式将返回 true。

2. containsIgnoreCase 函数

该函数与 contains 函数功能相似，但判断是不区分大写的。其定义如下：

`containsIgnoreCase(string, substring)`

该函数调用示例代码如下：

`${fn:containsIgnoreCase("I AM studying", "am") }`

上述 EL 表达式将返回 true。

3. endsWith 函数

该函数功能是判断一个字符串是否以指定的后缀结尾。其定义如下：

`endsWith(string, suffix)`

该函数调用示例代码如下：

`${fn:endsWith("I AM studying", "am") }`

上述 EL 表达式将返回 false。

4. indexOf 函数

该函数功能是返回指定子字符串在某个字符串中第一次出现时的索引，找不到时，将返回−1。其定义如下：

`indexOf(string, substring)`

该函数调用示例代码如下：

`${fn:indexOf("I am studying", "am") }`

上述 EL 表达式将返回 2。

5. join 函数

该函数功能是将一个 String 数组中的所有元素合并成一个字符串，并用指定的分隔符分开。其定义如下：

`join(array, separator)`

例如，假设一个 String 数组 my，它有 3 个元素："I""am"和"studying"，那么，下列 EL 表达式：

```
${fn:join(my, ",") }
```

将返回"I,am,studying"。

6. length 函数

该函数功能是返回集合中元素的个数,或者字符串中的字符个数。其定义如下:

```
length(input)
```

该函数调用示例代码如下:

```
${fn:length("aaa") }
```

上述 EL 表达式将返回 3。

7. replace 函数

该函数功能是将字符串中出现的所有 beforestring 用 afterstring 替换,并返回替换后的结果。其定义如下:

```
replace(string, beforestring, afterstring)
```

该函数调用示例代码如下:

```
${fn:replace("I am am studying", "am", "do") }
```

上述 EL 表达式将返回"I do do studying"。

8. split 函数

该函数功能是将一个字符串,使用指定的分隔符 separator 分离成一个子字符串数组。其定义如下:

```
split(string, separator)
```

该函数调用示例代码如下:

```
<c:set var="my" value="${fn:split('I am studying', ' ') }"/>
<c:forEach var="myArrayElement" items="${my }">
    ${myArrayElement}<br>
</c:forEach>
```

上述示例代码显示结果如图 11.5 所示。

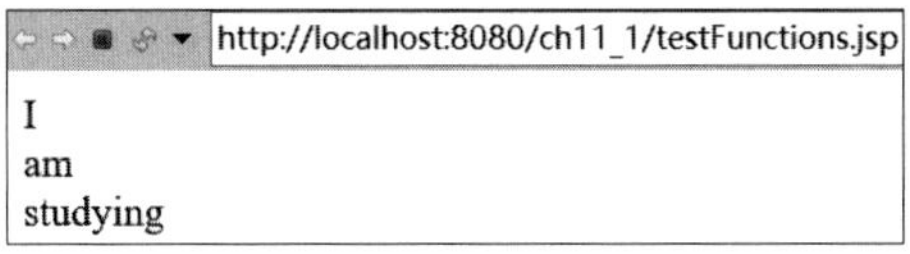

图 11.5　split 示例结果

9. startsWith 函数

该函数功能是判断一个字符串是否以指定的前缀开头。其定义如下：

startsWith(string, prefix)

该函数调用示例代码如下：

${fn:startsWith("I AM studying", "am") }

上述 EL 表达式将返回 false。

10. substring 函数

该函数功能是返回一个字符串的子字符串。其定义如下：

substring(string, begin, end)

该函数调用示例代码如下：

${fn:substring("abcdef", 1, 3) }

上述 EL 表达式将返回"bc"。

11. toLowerCase 函数

该函数功能是将一个字符串转换成它的小写版本。其定义如下：

toLowerCase(string)

该函数调用示例代码如下：

${fn:toLowerCase("I AM studying") }

上述 EL 表达式将返回"i am studying"。

12. toUpperCase 函数

该函数功能与 toLowerCase 函数的功能相反，不再赘述。

13. trim 函数

该函数功能是将一个字符串开头和结尾的空白去掉。其定义如下：

trim(string)

该函数调用示例代码如下：

${fn:trim("   I AM studying   ") }

上述 EL 表达式将返回"I AM studying"。

11.2.6 实践环节

编写一个 JSP 页面，在该页面中使用<c:forEach>标签输出九九乘法表。页面运行效果如图 11.6 所示。

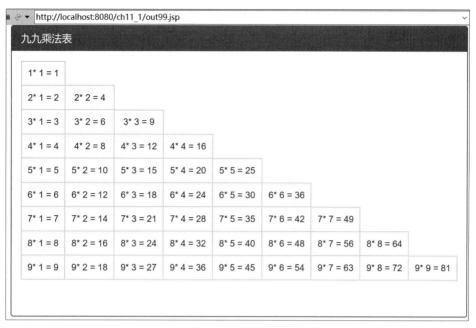

图 11.6　使用＜c:forEach＞打印九九乘法表

11.3　本 章 小 结

本章重点介绍了 EL 表达式、JSTL 核心标签库以及 JSTL 函数标签库的用法。EL 与 JSTL 的应用大大提高了编程效率,并且降低了维护难度。

习　题　11

1. 在 Web 应用程序中有以下程序代码段,执行后转到某个 JSP 页面：

```
ArrayList<String>  dogNames =new ArrayList<String>();
dogNames.add("goodDog");
request.setAttribute("dogs", dogNames);
```

以下( )选项可以正确地使用 EL 取得数组中的值。

　　A. ＄{ dogs .0}　　　　　　　　B. ＄{ dogs [0]}
　　C. ＄{ dogs .[0]}　　　　　　　　D. ＄{ dogs "0"}

2. ( )JSTL 标签可以实现 Java 程序中的 if 语句功能。

　　A. ＜c:set＞　　　　　　　　B. ＜c:out＞
　　C. ＜c:forEach＞　　　　　　D. ＜c:if＞

3. ( )不是 EL 的隐含对象。

A. request　　　　　　　　　　B. pageScope
　　C. sessionScope　　　　　　　 D. applicationScope
4.（　　）JSTL 标签可以实现 Java 程序中的 for 语句功能。
　　A. ＜c:set＞　　　　　　　　　B. ＜c:out＞
　　C. ＜c:forEach＞　　　　　　　D. ＜c:if＞

MyBatis 入门

学习目的与要求

本章讲解 MyBatis 的环境构建、工作原理以及与 Spring MVC 框架的整合开发。通过本章的学习,了解 MyBatis 的工作原理,掌握 MyBatis 的环境构建以及与 Spring MVC 框架的整合开发。

本章主要内容

- MyBatis 的环境构建。
- MyBatis 的工作原理。
- MyBatis 与 Spring MVC 框架整合开发。

MyBatis 是主流的 Java 持久层框架之一,它与 Hibernate 一样,也是一种对象关系映射(Object/Relational Mapping,ORM)框架。因其性能优异,且具有高度的灵活性、可优化性、易维护以及简单易学等特点,受到了广大互联网企业和编程爱好者的青睐。

12.1 MyBatis 简介

MyBatis 本是 apache 的一个开源项目 iBatis。2010 年，这个项目由 apache software foundation 迁移到 google code，并改名为 MyBatis。

MyBatis 是一个基于 Java 的持久层框架。MyBatis 提供的持久层框架包括 SQL Maps 和 Data Access Objects（DAO），它消除了几乎所有的 JDBC 代码和参数的手工设置以及结果集的检索。MyBatis 使用简单的 XML 或注解用于配置和原始映射，将接口和 Java 的普通的 Java 对象（Plain Old Java Objects，POJOs）映射成数据库中的记录。

目前，Java 的持久层框架产品有许多，常见的有 Hibernate 和 MyBatis。MyBatis 是一个半自动映射的框架，因为 MyBatis 需要手动匹配 POJO、SQL 和映射关系；而 Hibernate 是一个全表映射的框架，只需提供 POJO 和映射关系即可。MyBatis 是一个小巧、方便、高效、简单、直接、半自动化的持久层框架；Hibernate 是一个强大、方便、高效、复杂、间接、全自动化的持久化框架。两个持久层框架各有优缺点，开发者应根据实际应用选择它们。

12.2 MyBatis 的环境构建

编写本书时，MyBatis 的最新版本是 3.5.4，因此笔者选择这个版本作为本书的实践环境。也希望读者下载该版本，以便学习。

12.2.1 非 Maven 构建

如果读者不使用 Maven 或 Gradle 下载 MyBatis，可通过网址 https://github.com/mybatis/mybatis-3/releases 下载。下载时只需选择 mybatis-3.5.4.zip 即可，解压后得到图 12.1 所示的目录。

图 12.1 MyBatis 的目录

图 12.1 中的 mybatis-3.5.4.jar 是 MyBatis 的核心包，mybatis-3.5.4.pdf 是 MyBatis 的使用手册，lib 文件夹下的 JAR 是 MyBatis 的依赖包。

使用 MyBatis 框架时，需要将它的核心包和依赖包引入到应用程序中。如果是 Web 应用，只需将核心包和依赖包复制到/WEB-INF/lib 目录中。

12.2.2 Maven 构建

如果通过 Maven 下载 MyBatis，在应用程序的 pom.xml 文件中添加 MyBatis 的核心包即可。这是因为 Maven 将自动添加 MyBatis 的依赖包。添加 MyBatis 的核心包代码如下：

```xml
<dependency>
    <groupId>org.mybatis</groupId>
    <artifactId>mybatis</artifactId>
    <version>3.5.4</version>
</dependency>
```

12.3　MyBatis 的工作原理

学习 MyBatis 程序之前，需要了解一下 MyBatis 的工作原理，以便理解程序。MyBatis 的工作原理如图 12.2 所示。

下面对图 12.2 中的每步流程进行说明，具体如下。

（1）读取 MyBatis 配置文件 mybatis-config.xml。mybatis-config.xml 为 MyBatis 的全局配置文件，配置了 MyBatis 的运行环境等信息，如数据库连接信息。

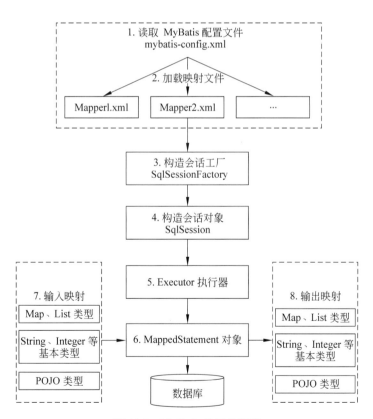

图 12.2　MyBatis 的工作原理

（2）加载映射文件。映射文件即 SQL 映射文件，文件中配置了操作数据库的 SQL 语句，需要在 MyBatis 配置文件 mybatis-config.xml 中加载。mybatis-config.xml 文件可以加载多个映射文件。

（3）构造会话工厂。通过 MyBatis 的环境等配置信息构建会话工厂 SqlSessionFactory。

（4）创建 SqlSession 对象。由会话工厂创建 SqlSession 对象，该对象中包含执行 SQL 语句的所有方法。

（5）MyBatis 底层定义了一个 Executor 接口来操作数据库，它将根据 SqlSession 传递的参数动态地生成需要执行的 SQL 语句，同时负责查询缓存的维护。

（6）在 Executor 接口的执行方法中，有一个 MappedStatement 类型的参数，该参数是对映射信息的封装，用于存储要映射的 SQL 语句的 id、参数等信息。

（7）输入参数映射。输入参数类型既可以是 Map、List 等集合类型，也可以是基本数据类型和 POJO 类型。输入参数映射过程类似 JDBC 对 preparedStatement 对象设置参数的过程。

（8）输出结果映射。输出结果类型既可以是 Map、List 等集合类型，也可以是基本数据类型和 POJO 类型。输出结果映射过程类似 JDBC 对结果集的解析过程。

通过上面的讲解，读者对 MyBatis 框架应该有一个初步了解。在后续的学习中再慢慢加深理解。

12.4 使用 STS 开发 MyBatis 入门程序

本节使用第 1.7.3 节的 MySQL 数据库 springtest 的 user 数据表讲解。下面通过一个实例讲解如何使用 STS 开发 MyBatis 入门程序。

【例 12-1】 使用 STS 开发 MyBatis 入门程序。

具体实现步骤如下：

12.4.1 创建 Maven 项目并添加相关依赖

使用 STS 创建一个名为 ch12_1 的 Maven Project，并通过 pom.xml 文件添加项目所依赖的 JAR 包。该实例中除了添加 MyBatis 的核心依赖外，还需要添加 MySQL 连接依赖及 Log4j 日志依赖。ch12_1 的 pom.xml 文件内容如下：

```
<project xmlns="http://maven.apache.org/POM/4.0.0"
    xmlns:xsi="http://www.w3.org/2001/XMLSchema-instance"
    xsi:schemaLocation="http://maven.apache.org/POM/4.0.0
https://maven.apache.org/xsd/maven-4.0.0.xsd">
    <modelVersion>4.0.0</modelVersion>
    <groupId>com.cn.chenheng</groupId>
    <artifactId>ch12_1</artifactId>
    <version>0.0.1-SNAPSHOT</version>
    <packaging>war</packaging>
    <dependencies>
```

```xml
<dependency>
    <groupId>org.mybatis</groupId>
    <artifactId>mybatis</artifactId>
    <version>3.5.4</version>
</dependency>
<dependency>
    <groupId>mysql</groupId>
    <artifactId>mysql-connector-java</artifactId>
    <version>5.1.45</version>
</dependency>
<dependency>
    <groupId>log4j</groupId>
    <artifactId>log4j</artifactId>
    <version>1.2.17</version>
</dependency>
    </dependencies>
</project>
```

12.4.2 创建 Log4j 的日志配置文件

MyBatis 可使用 Log4j 输出日志信息，如果开发者需要查看控制台输出的 SQL 语句，则需要在 classpath 路径下配置其日志文件。在应用 ch12_1 的 src/main/resources 目录下创建 log4j.properties 文件，其内容如下：

```
#Global logging configuration
log4j.rootLogger=ERROR, stdout
#MyBatis logging configuration...
log4j.logger.com.mybatis=DEBUG
#Console output...
log4j.appender.stdout=org.apache.log4j.ConsoleAppender
log4j.appender.stdout.layout=org.apache.log4j.PatternLayout
log4j.appender.stdout.layout.ConversionPattern=%5p [%t] - %m%n
```

上述日志文件中配置了全局的日志配置、MyBatis 的日志配置和控制台输出，其中 MyBatis 的日志配置用于将 com.mybatis 包下所有类的日志记录级别设置为 DEBUG。该配置文件内容不需要开发者全部手写，可以从 MyBatis 使用手册中 Logging 小节复制，然后进行简单修改。

Log4j 是 Apache 的一个开源代码项目，通过使用 Log4j，可以控制日志信息输送的目的地是控制台、文件或 GUI 组件等；也可以控制每一条日志的输出格式；通过定义每一条日志信息的级别，能够更加详细地控制日志的生成过程。这些都可以通过一个配置文件来灵活地进行配置，而不需要修改应用的代码。有关 Log4j 的使用方法，读者可参考相关资料学习。

12.4.3 创建持久化类

在应用 ch12_1 的 src/main/java 目录下创建一个名为 com.mybatis.po 的包,并在该包中创建持久化类 MyUser。类中声明的属性与数据表 user(创建表的代码请参见源代码的 user.sql)的字段一致。

MyUser 类的代码如下:

```
package com.mybatis.po;
/**
 *springtest 数据库中 user 表的持久化类
 */
public class MyUser {
    private Integer uid;//主键
    private String uname;
    private String usex;
    //此处省略 setter 和 getter 方法
    @Override
    public String toString() {//为了方便查看结果,重写了 toString 方法
        return "User [uid=" +uid +",uname=" +uname +",usex=" +usex +"]";
    }
}
```

12.4.4 创建 SQL 映射文件

在应用 ch12_1 的 src/main/resources 目录下创建一个名为 com.mybatis.mapper 的包,并在该包中创建 SQL 映射文件 UserMapper.xml。

SQL 映射文件 UserMapper.xml 文件内容如下:

```
<?xml version="1.0" encoding="UTF-8" ?>
<!DOCTYPE mapper
PUBLIC "-//mybatis.org//DTD Mapper 3.0//EN"
"http://mybatis.org/dtd/mybatis-3-mapper.dtd">
<mapper namespace="com.mybatis.mapper.UserMapper">
    <!--根据 uid 查询一个用户信息 -->
    <select id="selectUserById" parameterType="Integer"
        resultType="com.mybatis.po.MyUser">
        select * from user where uid =#{uid}
    </select>
    <!--查询所有用户信息 -->
    <select id="selectAllUser"  resultType="com.mybatis.po.MyUser">
        select * from user
    </select>
    <!--添加一个用户 ,#{uname}为 com.mybatis.po.MyUser 的属性值-->
    <insert id="addUser" parameterType="com.mybatis.po.MyUser">
```

```
        insert into user (uname,usex) values(#{uname},#{usex})
    </insert>
    <!--修改一个用户-->
    <update id="updateUser" parameterType="com.mybatis.po.MyUser">
        update user set uname=#{uname},usex=#{usex} where uid=#{uid}
    </update>
    <!--删除一个用户-->
    <delete id="deleteUser" parameterType="Integer">
        delete from user where uid=#{uid}
    </delete>
</mapper>
```

在上述映射文件中，<mapper>元素是配置文件的根元素，它包含了一个 namespace 属性，该属性值通常设置为"包名＋SQL 映射文件名"，指定了唯一的命名空间。子元素 <select><insert><update>以及<delete>中的信息是用于执行查询、添加、修改以及删除操作的配置。在定义的 SQL 语句中，"♯{}"表示一个占位符，相当于"？"，而"♯{uid}"表示该占位符待接收参数的名称为 uid。

12.4.5 创建 MyBatis 的核心配置文件

在应用 ch12_1 的 src/main/resources 目录下创建 MyBatis 的核心配置文件 mybatis-config.xml。该文件中配置了数据库环境和映射文件的位置，具体内容如下：

```
<?xml version="1.0" encoding="UTF-8" ?>
<!DOCTYPE configuration
PUBLIC "-//mybatis.org//DTD Config 3.0//EN"
"http://mybatis.org/dtd/mybatis-3-config.dtd">
<configuration>
    <!--配置环境-->
    <environments default="development">
        <environment id="development">
            <!--使用 JDBC 的事务管理-->
            <transactionManager type="JDBC"/>
            <dataSource type="POOLED">
                <!--MySQL 数据库驱动-->
                <property name="driver" value="com.mysql.jdbc.Driver"/>
                <!--连接数据库的 URL-->
                <property name="url"
                    value="jdbc:mysql://localhost:3306/
                    springtest?characterEncoding=utf8"/>
                <property name="username" value="root"/>
                <property name="password" value="root"/>
            </dataSource>
        </environment>
    </environments>
```

```xml
<mappers>
<!--映射文件的位置 -->
<mapper resource="com/mybatis/mapper/UserMapper.xml"/>
</mappers>
</configuration>
```

上述映射文件和配置文件都不需要读者完全手动编写,都可以从 MyBatis 使用手册中复制,然后做简单修改即可。

12.4.6 创建测试类

在应用 ch12_1 的 src/main/java 目录下创建一个名为 com.mybatis.test 的包,并在该包中创建 MyBatisTest 测试类。在测试类中,首先使用输入流读取配置文件,然后根据配置信息构建 SqlSessionFactory 对象。接下来通过 SqlSessionFactory 对象创建 SqlSession 对象,并使用 SqlSession 对象执行数据库操作。

MyBatisTest 测试类的代码如下:

```java
package com.mybatis.test;
import java.io.IOException;
import java.io.InputStream;
import java.util.List;
import org.apache.ibatis.io.Resources;
import org.apache.ibatis.session.SqlSession;
import org.apache.ibatis.session.SqlSessionFactory;
import org.apache.ibatis.session.SqlSessionFactoryBuilder;
import com.mybatis.po.MyUser;
public class MyBatisTest {
    public static void main(String[] args) {
        try {
            //读取配置文件 mybatis-config.xml
            InputStream config = Resources.getResourceAsStream("mybatis-config.xml");
            //根据配置文件构建 SqlSessionFactory
            SqlSessionFactory ssf = new SqlSessionFactoryBuilder().build(config);
            //通过 SqlSessionFactory 创建 SqlSession
            SqlSession ss = ssf.openSession();
            //SqlSession 执行映射文件中定义的 SQL,并返回映射结果
            /* com.mybatis.mapper.UserMapper.selectUserById 为
            UserMapper.xml 中的命名空间+select 的 id */
            //查询一个用户
            MyUser mu = ss.selectOne("com.mybatis.mapper.UserMapper.selectUserById", 1);
            System.out.println(mu);
            //添加一个用户
            MyUser addmu = new MyUser();
            addmu.setUname("陈恒");
```

```
            addmu.setUsex("男");
            ss.insert("com.mybatis.mapper.UserMapper.addUser",addmu);
            //修改一个用户
            MyUser updatemu =new MyUser();
            updatemu.setUid(1);
            updatemu.setUname("张三");
            updatemu.setUsex("女");
            ss.update("com.mybatis.mapper.UserMapper.updateUser", updatemu);
            //删除一个用户
            ss.delete("com.mybatis.mapper.UserMapper.deleteUser", 3);
            //查询所有用户
            List<MyUser>listMu = ss.selectList("com.mybatis.mapper.UserMapper.selectAllUser");
            for (MyUser myUser : listMu) {
                System.out.println(myUser);
            }
            //提交事务
            ss.commit();
            //关闭 SqlSession
            ss.close();
        } catch (IOException e) {
            // TODO Auto-generated catch block
            e.printStackTrace();
        }
    }
}
```

上述测试类的运行结果如图 12.3 所示。

```
DEBUG [main] - ==>  Preparing: select * from user where uid = ?
DEBUG [main] - ==> Parameters: 1(Integer)
DEBUG [main] - <==      Total: 0
null
DEBUG [main] - ==>  Preparing: insert into user (uname,usex) values(?,?)
DEBUG [main] - ==> Parameters: 陈恒(String), 男(String)
DEBUG [main] - <==    Updates: 1
DEBUG [main] - ==>  Preparing: update user set uname = ?,usex = ? where uid = ?
DEBUG [main] - ==> Parameters: 张三(String), 女(String), 1(Integer)
DEBUG [main] - <==    Updates: 1
DEBUG [main] - ==>  Preparing: delete from user where uid = ?
DEBUG [main] - ==> Parameters: 3(Integer)
DEBUG [main] - <==    Updates: 0
DEBUG [main] - ==>  Preparing: select * from user
DEBUG [main] - ==> Parameters:
DEBUG [main] - <==      Total: 1
User [uid=1,uname=张三,usex=女]
```

图 12.3 MyBatis 入门程序的运行结果

12.5　MyBatis 与 Spring MVC 的整合开发

从 12.4 节测试类的代码中，可以看出直接使用 MyBatis 框架的 SqlSession 访问数据库并不简便。MyBatis 框架的重点是 SQL 映射文件，因此为方便后续学习，本节开始讲解 MyBatis 与 Spring MVC 的整合。在本书 MyBatis 的后续学习中，将使用整合后的框架进行演示。

12.5.1　相关依赖

实现 MyBatis 与 Spring MVC 的整合，需要在应用的 pom.xml 文件中添加相关依赖，包括 MyBatis、Spring MVC、Spring JDBC、MySQL 连接器、MyBatis 与 Spring 桥接器、Log4j 以及 DBCP 等依赖。

1. MyBatis 依赖

在应用的 pom.xml 文件中添加 MyBatis 依赖的代码如下：

```
<dependency>
    <groupId>org.mybatis</groupId>
    <artifactId>mybatis</artifactId>
    <version>3.5.4</version>
</dependency>
```

2. Spring MVC 依赖

在应用的 pom.xml 文件中添加 Spring MVC 依赖的代码如下：

```
<dependency>
    <groupId>org.springframework</groupId>
    <artifactId>spring-webmvc</artifactId>
    <version>${spring.version}</version>
</dependency>
```

3. Spring JDBC 依赖

在应用的 pom.xml 文件中添加 Spring JDBC 依赖的代码如下：

```
<dependency>
    <groupId>org.springframework</groupId>
    <artifactId>spring-jdbc</artifactId>
    <version>${spring.version}</version>
</dependency>
```

4. MySQL 连接器

在应用的 pom.xml 文件中添加 MySQL 连接器依赖的代码如下：

```xml
<dependency>
    <groupId>mysql</groupId>
    <artifactId>mysql-connector-java</artifactId>
    <version>5.1.45</version>
</dependency>
```

5. MyBatis 与 Spring 桥接器依赖

MyBatis 与 Spring 桥接器将 MyBatis 代码无缝地整合到 Spring 框架中。它允许 MyBatis 参与到 Spring 框架的事务管理中，创建映射器 Mapper 和 SqlSession 并注入 Spring 容器中，以及将 MyBatis 的异常转换为 Spring 的 DataAccessException。在应用的 pom.xml 文件中添加 MyBatis 与 Spring 桥接器（编写本书时，最新版是 2.0.3）依赖的代码如下：

```xml
<dependency>
    <groupId>org.mybatis</groupId>
    <artifactId>mybatis-spring</artifactId>
    <version>2.0.3</version>
</dependency>
```

6. Log4j 依赖

MyBatis 与 Spring MVC 整合时，可使用 Log4j 输出日志信息。在应用的 pom.xml 文件中添加 Log4j（编写本书时，最新版是 1.2.17）依赖的代码如下：

```xml
<dependency>
    <groupId>log4j</groupId>
    <artifactId>log4j</artifactId>
    <version>1.2.17</version>
</dependency>
```

7. DBCP 依赖

MyBatis 与 Spring MVC 整合时，将使用 DBCP 数据源，需要准备 DBCP（commons-dbcp2）和连接池（commons-pool2）的相关依赖。而 commons-pool2 依赖 commons-dbcp2，所以在应用的 pom.xml 文件中添加 commons-dbcp2（编写本书时，最新版是 2.7.0）依赖即可，具体代码如下：

```xml
<dependency>
    <groupId>org.apache.commons</groupId>
    <artifactId>commons-dbcp2</artifactId>
    <version>2.7.0</version>
</dependency>
```

12.5.2 在 Sping MVC 的配置类中配置数据源及 MyBatis 工厂

在一般情况下，将数据源及 MyBatis 工厂配置在 Spring MVC 的 Java 配置类中，实现 MyBatis 与 Spring MVC 的无缝整合。

在 Spring MVC 的配置类中，首先使用 org.apache.commons.dbcp2.BasicDataSource 配置数据源；其次使用 org.springframework.jdbc.datasource.DataSourceTransactionManager 为数据源添加事务管理器；再次使用 org.mybatis.spring.SqlSessionFactoryBean 配置 MyBatis 工厂，同时指定数据源，并与 MyBatis 完美整合；最后使用 org.mybatis.spring.annotation.MapperScan 扫描数据访问层，将数据访问层中的所有接口自动装配为 MyBatis 的 Mapper 接口的实现类。配置类的示例代码如下：

```java
package config;
import org.apache.commons.dbcp2.BasicDataSource;
import org.mybatis.spring.SqlSessionFactoryBean;
import org.mybatis.spring.annotation.MapperScan;
import org.springframework.beans.factory.annotation.Value;
import org.springframework.context.annotation.Bean;
import org.springframework.context.annotation.ComponentScan;
import org.springframework.context.annotation.Configuration;
import org.springframework.context.annotation.PropertySource;
import org.springframework.core.io.ClassPathResource;
import org.springframework.core.io.Resource;
import org.springframework.jdbc.datasource.DataSourceTransactionManager;
import org.springframework.transaction.annotation.EnableTransactionManagement;
import org.springframework.web.servlet.config.annotation.EnableWebMvc;
import org.springframework.web.servlet.view.InternalResourceViewResolver;
@Configuration  // 通过该注解表明该类是一个 Spring 的配置，相当于一个 xml 文件
@EnableWebMvc  // 开启 Spring MVC 的支持
@ComponentScan(basePackages = { "controller"})  // 扫描基本包
@PropertySource(value = { "classpath:jdbc.properties" }, ignoreResourceNotFound = true)
//配置多个配置文件  value={"classpath:jdbc.properties","xx","xxx"}
@EnableTransactionManagement  // 开启声明式事务的支持
@MapperScan(basePackages = { "com.mybatis.mapper" })  // 配置扫描 Mapper 接口的包路径
public class SpringJDBCConfig {
    @Value("${db.url}")  // 注入属性文件 jdbc.properties 中的 jdbc.url
    private String jdbcUrl;
    @Value("${db.driverClassName}")
    private String jdbcDriverClassName;
    @Value("${db.username}")
    private String jdbcUsername;
    @Value("${db.password}")
    private String jdbcPassword;
```

```java
@Value("${db.maxTotal}")
private int maxTotal;
@Value("${db.maxIdle}")
private int maxIdle;
@Value("${db.initialSize}")
private int initialSize;
/**
 * 配置数据源
 */
@Bean
public BasicDataSource dataSource() {
    BasicDataSource myDataSource = new BasicDataSource();
    // 数据库驱动
    myDataSource.setDriverClassName(jdbcDriverClassName);
    // 相应驱动的 jdbcUrl
    myDataSource.setUrl(jdbcUrl);
    // 数据库的用户名
    myDataSource.setUsername(jdbcUsername);
    // 数据库的密码
    myDataSource.setPassword(jdbcPassword);
    // 最大连接数
    myDataSource.setMaxTotal(maxTotal);
    // 最大空闲连接数
    myDataSource.setMaxIdle(maxIdle);
    // 初始化连接数
    myDataSource.setInitialSize(initialSize);
    return myDataSource;
}
/**
 * 为数据源添加事务管理器
 */
@Bean
public DataSourceTransactionManager transactionManager() {
    DataSourceTransactionManager dt = new DataSourceTransactionManager();
    dt.setDataSource(dataSource());
    return dt;
}
/**
 * 配置 MyBatis 工厂,同时指定数据源,并与 MyBatis 完美整合
 */
@Bean
public SqlSessionFactoryBean getSqlSession() {
    SqlSessionFactoryBean sqlSessionFactory = new SqlSessionFactoryBean();
    sqlSessionFactory.setDataSource(dataSource());
```

```
        Resource r =new ClassPathResource("mybatis-config.xml");
        sqlSessionFactory.setConfigLocation(r);
        return sqlSessionFactory;
    }
    /**
     * 配置视图解析器
     */
    @Bean
    public InternalResourceViewResolver getViewResolver() {
        InternalResourceViewResolver viewResolver =new 
        InternalResourceViewResolver();
        viewResolver.setPrefix("/WEB-INF/jsp/");
        viewResolver.setSuffix(".jsp");
        return viewResolver;
    }
}
```

12.5.3 整合示例

下面通过 MyBatis 与 Spring MVC 的整合实现【例 12-1】的功能。

【例 12-2】 MyBatis 与 Spring MVC 的整合开发。

具体实现步骤如下。

1. 使用 STS 创建 Maven 项目并添加相关依赖

使用 STS 创建一个名为 ch12_2 的 Maven Project，并参考 12.5.1 节，通过 pom.xml 文件添加项目所依赖的 JAR 包。ch12_2 的 pom.xml 文件内容如下：

```xml
<project xmlns="http://maven.apache.org/POM/4.0.0"
  xmlns:xsi="http://www.w3.org/2001/XMLSchema-instance"
  xsi:schemaLocation=" http://maven. apache. org/POM/4.0.0  https://maven.
apache.org/xsd/maven-4.0.0.xsd">
  <modelVersion>4.0.0</modelVersion>
  <groupId>com.cn.chenheng</groupId>
  <artifactId>ch12_2</artifactId>
  <version>0.0.1-SNAPSHOT</version>
  <packaging>war</packaging>
  <properties>
      <!--spring版本号 -->
      <spring.version>5.2.3.RELEASE</spring.version>
  </properties>
  <dependencies>
      <dependency>
          <groupId>org.springframework</groupId>
          <artifactId>spring-webmvc</artifactId>
```

```xml
            <version>${spring.version}</version>
        </dependency>
        <dependency>
            <groupId>org.springframework</groupId>
            <artifactId>spring-jdbc</artifactId>
            <version>${spring.version}</version>
        </dependency>
        <dependency>
            <groupId>mysql</groupId>
            <artifactId>mysql-connector-java</artifactId>
            <version>5.1.45</version>
        </dependency>
        <dependency>
            <groupId>org.apache.commons</groupId>
            <artifactId>commons-dbcp2</artifactId>
            <version>2.7.0</version>
        </dependency>
        <dependency>
            <groupId>org.mybatis</groupId>
            <artifactId>mybatis</artifactId>
            <version>3.5.4</version>
        </dependency>
        <dependency>
            <groupId>org.mybatis</groupId>
            <artifactId>mybatis-spring</artifactId>
            <version>2.0.3</version>
        </dependency>
        <dependency>
            <groupId>log4j</groupId>
            <artifactId>log4j</artifactId>
            <version>1.2.17</version>
        </dependency>
    </dependencies>
</project>
```

2. 创建数据库连接信息属性文件及 Log4j 的日志配置文件

在应用 ch12_2 的 src/main/resources 目录下创建数据库连接信息属性文件 jdbc.properties 文件，具体内容如下：

```
db.driverClassName=com.mysql.jdbc.Driver
db.url=jdbc:mysql://localhost:3306/springtest?characterEncoding=utf8
db.username=root
db.password=root
```

```
db.maxTotal=30
db.maxIdle=15
db.initialSize=5
```

在应用 ch12_2 的 src/main/resources 目录下创建 Log4j 的日志配置文件 log4j.properties 文件，其内容与 12.4.2 节相同，不再赘述。

3. 创建持久化类

在应用 ch12_2 的 src/main/java 目录下创建一个名为 com.mybatis.po 的包，并在该包中创建持久化类 MyUser。该类与 12.4.3 节相同，不再赘述。

4. 创建 SQL 映射文件

在应用 ch12_2 的 src/main/resources 目录下创建一个名为 com.mybatis.mapper 的包，并在该包中创建 SQL 映射文件 UserMapper.xml。该文件与 12.4.4 节相同，不再赘述。

5. 创建 MyBatis 的核心配置文件

在应用 ch12_2 的 src/main/resources 目录下创建 MyBatis 的核心配置文件 mybatis-config.xml。在该文件中指定 SQL 映射文件的位置，而数据源的配置在 Spring MVC 的 Java 配置类中完成。mybatis-config.xml 的内容如下：

```xml
<?xml version="1.0" encoding="UTF-8" ?>
<!DOCTYPE configuration
PUBLIC "-//mybatis.org//DTD Config 3.0//EN"
"http://mybatis.org/dtd/mybatis-3-config.dtd">
<configuration>
    <settings>
        <!--输出日志配置 -->
        <setting name="logImpl" value="LOG4J" />
    </settings>
    <mappers>
        <!--映射文件的位置 -->
        <mapper resource="com/mybatis/mapper/UserMapper.xml"/>
    </mappers>
</configuration>
```

6. 创建 Mapper 接口

在应用 ch12_2 的 src/main/java 目录下创建一个名为 com.mybatis.mapper 的包，并在该包中创建 Mapper 接口 UserMapper。该接口中的方法与 com.mybatis.mapper 中的 SQL 映射文件 UserMapper.xml 的 id 一致。UserMapper 接口代码如下：

```
package com.mybatis.mapper;
import java.util.List;
import org.springframework.stereotype.Repository;
import com.mybatis.po.MyUser;
@Repository
public interface UserMapper {
    public MyUser selectUserById(Integer id);
    public List<MyUser>selectAllUser();
    public int addUser(MyUser myUser);
    public int updateUser(MyUser myUser);
    public int deleteUser(Integer id);
}
```

7. 创建控制类

在应用 ch12_2 的 src/main/java 目录下创建一个名为 controller 的包，并在该包中创建控制器类 TestController。在该控制器类中调用 Mapper 接口中的方法操作数据库，具体代码如下：

```
package controller;
import java.util.List;
import org.springframework.beans.factory.annotation.Autowired;
import org.springframework.stereotype.Controller;
import org.springframework.web.bind.annotation.RequestMapping;
import com.mybatis.mapper.UserMapper;
import com.mybatis.po.MyUser;
@Controller
public class TestController {
@Autowired
private UserMapper userMapper;
@RequestMapping("/test")
public String test() {
    //查询一个用户
    MyUser mu =userMapper.selectUserById(1);
    System.out.println(mu);
    //添加一个用户
    MyUser addmu =new MyUser();
    addmu.setUname("陈恒");
    addmu.setUsex("男");
    userMapper.addUser(addmu);
    //修改一个用户
    MyUser updatemu =new MyUser();
```

```
            updatemu.setUid(1);
            updatemu.setUname("张三");
            updatemu.setUsex("女");
            userMapper.updateUser(updatemu);
            //删除一个用户
            userMapper.deleteUser(3);
            //查询所有用户
            List<MyUser> listMu = userMapper.selectAllUser();
            for (MyUser myUser : listMu) {
                System.out.println(myUser);
            }
            return "test";
        }
    }
```

8. 创建测试页面

在应用 ch12_2 的 src/main/webapp/WEB-INF/目录下创建一个名为 jsp 的文件夹，并在该文件夹中创建 test.jsp 文件，具体代码如下：

```
<%@ page language="java" contentType="text/html; charset=UTF-8" pageEncoding="UTF-8"%>
<%
String path = request.getContextPath();
String basePath = request.getScheme() + "://" + request.getServerName() + ":" + request.getServerPort()+path+"/";
%>
<!DOCTYPE html>
<html>
<head>
<base href="<%=basePath%>">
<meta charset="UTF-8">
<title>Insert title here</title>
</head>
<body>
测试成功。
</body>
</html>
```

9. 创建 Web 和 Spring MVC 配置类

在应用 ch12_2 的 src/main/java 目录下创建一个名为 config 的包，并在该包中创建 Web 配置类 WebConfig 和 Spring MVC 的配置类 SpringJDBCConfig。

在 Spring MVC 的配置类 SpringJDBCConfig 中，首先使用 BasicDataSource 配置数

据源；其次使用 DataSourceTransactionManager 为数据源添加事务管理器；最后使用 SqlSessionFactoryBean 配置 MyBatis 工厂，同时指定数据源，并与 MyBatis 完美整合；最后使用 MapperScan 扫描数据访问层，将数据访问层中的所有接口自动装配为 MyBatis 的 Mapper 接口的实现类。SpringJDBCConfig 的代码与 12.5.2 节相同，不再赘述。

在 Web 配置类 WebConfig 中注册配置类 SpringJDBCConfig，具体代码如下：

```
package config;
import javax.servlet.ServletContext;
import javax.servlet.ServletException;
import javax.servlet.ServletRegistration.Dynamic;
import org.springframework.web.WebApplicationInitializer;
import org.springframework.web.context.support.
AnnotationConfigWebApplicationContext;
import org.springframework.web.servlet.DispatcherServlet;
public class WebConfig implements WebApplicationInitializer{
    @Override
    public void onStartup(ServletContext arg0) throws ServletException {
        AnnotationConfigWebApplicationContext ctx
        =new AnnotationConfigWebApplicationContext();
        ctx.register(SpringJDBCConfig.class);//注册配置类 SpringJDBCConfig
        ctx.setServletContext(arg0);//和当前 ServletContext 关联
        /**
         * 注册 Spring MVC 的 DispatcherServlet
         */
        Dynamic servlet =arg0.addServlet("dispatcher", new DispatcherServlet(ctx));
        servlet.addMapping("/");
        servlet.setLoadOnStartup(1);
    }
}
```

10. 测试应用

发布应用 ch12_2 到 Web 服务器 Tomcat 后，通过地址 http://localhost:8080/ch12_2/test 测试应用。运行成功后，控制台信息输出结果，如图 12.4 所示。

12.5.4　实践环节

参考 12.5.3 节的整合示例的实现过程，实现陈姓用户信息的查询。

```
DEBUG [http-nio-8080-exec-3] - ==>  Preparing: select * from user where uid = ?
DEBUG [http-nio-8080-exec-3] - ==> Parameters: 1(Integer)
DEBUG [http-nio-8080-exec-3] - <==      Total: 1
User [uid=1,uname=张三,usex=女]
DEBUG [http-nio-8080-exec-3] - ==>  Preparing: insert into user (uname,usex) valu
DEBUG [http-nio-8080-exec-3] - ==> Parameters: 陈恒(String), 男(String)
DEBUG [http-nio-8080-exec-3] - <==    Updates: 1
DEBUG [http-nio-8080-exec-3] - ==>  Preparing: update user set uname = ?,usex = ?
DEBUG [http-nio-8080-exec-3] - ==> Parameters: 张三(String), 女(String), 1(Intege
DEBUG [http-nio-8080-exec-3] - <==    Updates: 1
DEBUG [http-nio-8080-exec-3] - ==>  Preparing: delete from user where uid = ?
DEBUG [http-nio-8080-exec-3] - ==> Parameters: 3(Integer)
DEBUG [http-nio-8080-exec-3] - <==    Updates: 0
DEBUG [http-nio-8080-exec-3] - ==>  Preparing: select * from user
DEBUG [http-nio-8080-exec-3] - ==> Parameters:
DEBUG [http-nio-8080-exec-3] - <==      Total: 2
User [uid=1,uname=张三,usex=女]
User [uid=2,uname=陈恒,usex=男]
```

图 12.4　应用 ch12_2 的控制台信息输出结果

12.6　使用 MyBatis Generator 插件自动生成映射文件

使用 MyBatis Generator 插件自动生成 MyBatis 的 DAO 接口、实体模型类、Mapper 映射文件，将生成的代码复制到项目工程中即可，把更多精力放在业务逻辑上。

MyBatis Generator 有三种常用方法自动生成代码：命令行、Eclipse 插件和 Maven 插件。本节使用比较简单的方法（命令行）自动生成相关代码。具体步骤如下：

1. 准备相关 JAR 包

需要准备的 JAR 包：mysql-connector-java-5.1.45-bin.jar 和 mybatis-generator-core-1.4.0.jar（https://mvnrepository.com/artifact/org.mybatis.generator/mybatis-generator-core/1.4.0）。

2. 创建文件目录

在某磁盘根目录下新建一个文件目录，如 C:\generator，并将 mysql-connector-java-5.1.45-bin.jar 和 mybatis-generator-core-1.4.0.jar 文件复制到 generator 目录下。另外，在 generator 目录下创建 src 子目录存放生成的相关代码文件。

3. 创建配置文件

在第二步创建的文件目录（C:\generator）下创建配置文件，如 C:\generator\generator.xml。文件目录如图 12.5 所示。

名称	修改日期
src	2020/2/15 7:41
generator.xml	2020/2/15 7:41
mybatis-generator-core-1.4.0.jar	2020/2/15 7:36
mysql-connector-java-5.1.45-bin.jar	2018/1/22 6:57

图 12.5　generator 目录

generator.xml 配置文件内容如下（具体含义见注释）：

```xml
<?xml version="1.0" encoding="UTF-8"?>
<!DOCTYPE generatorConfiguration PUBLIC " -//mybatis.org//DTD MyBatis Generator Configuration 1.0//EN" "http://mybatis.org/dtd/mybatis-generator-config_1_0.dtd">
<generatorConfiguration>
    <!--数据库驱动包位置 -->
    <classPathEntry location="C:\generator\mysql-connector-java-5.1.45-bin.jar" />
    <context id="mysqlTables" targetRuntime="MyBatis3">
        <commentGenerator>
            <property name="suppressAllComments" value="true" />
        </commentGenerator>
        <!--数据库链接 URL、用户名、密码（前提数据库 springtest 存在）-->
         <jdbcConnection
             driverClass="com.mysql.jdbc.Driver"
             connectionURL=" jdbc: mysql://localhost: 3306/springtest?characterEncoding=utf8"
             userId="root" password="root">
        </jdbcConnection>
        <javaTypeResolver>
            <property name="forceBigDecimals" value="false" />
        </javaTypeResolver>
        <!--生成模型（MyBatis 里面用到实体类）的包名和位置-->
        <javaModelGenerator targetPackage="com.po" targetProject="C:\generator\src">
            <property name="enableSubPackages" value="true" />
            <property name="trimStrings" value="true" />
        </javaModelGenerator>
        <!--生成的映射文件（MyBatis 的 SQL 语句 xml 文件）包名和位置-->
        <sqlMapGenerator targetPackage="mybatis" targetProject="C:\generator\src">
            <property name="enableSubPackages" value="true" />
        </sqlMapGenerator>
        <!--生成 DAO 的包名和位置 -->
        <javaClientGenerator type="XMLMAPPER" targetPackage="com.dao"
```

```
                targetProject="C:\generator\src">
                    <property name="enableSubPackages" value="true" />
            </javaClientGenerator>
            <!--要生成那些表（更改 tableName 和 domainObjectName 就可以，前提数据库
    springtest 中的 user 表已创建）-->
            <table tableName="user" domainObjectName="User" enableCountByExample=
    "false"   enableUpdateByExample = "false"   enableDeleteByExample = "false"
    enableSelectByExample="false" selectByExampleQueryId="false" />
        </context>
    </generatorConfiguration>
```

4. 使用命令生成代码

打开命令提示符，进入 C:\generator，输入命令 java -jar mybatis-generator-core-1.4.0.jar -configfile generator.xml – overwrite，如图 12.6 所示。

```
C:\>cd generator

C:\generator>java -jar mybatis-generator-core-1.4.0.jar -configfile generator.xml -overwrite
MyBatis Generator finished successfully.

C:\generator>
```

图 12.6　使用命令行生成映射文件

12.7　小　　结

本章重点讲述了 MyBatis 与 Spring MVC 框架的集成过程。通过本章的学习，读者不仅掌握 MyBatis 与 Spring MVC 框架的集成过程，还应该熟悉 MyBatis 的基本应用。

习　题　12

1. MyBatis Generator 有哪几种方法自动生成代码？
2. 简述 MyBatis 与 Spring MVC 框架集成的步骤。

MyBatis 的映射器

学习目的与要求

本章重点讲解 MyBatis 的 SQL 映射文件。通过本章的学习，了解 MyBatis 的核心配置文件的配置信息，掌握 MyBatis 的 SQL 映射文件的编写，熟悉级联查询的 MyBatis 实现，掌握 MyBatis 的动态 SQL 的编写方法。

本章主要内容

- 核心配置文件。
- SQL 映射文件。
- 级联查询。
- 动态 SQL。

MyBatis 框架的强大之处体现在 SQL 映射文件的编写上。因此，本章将重点讲解 SQL 映射文件的编写。

13.1 MyBatis 的核心配置

MyBatis 的核心配置文件配置了影响 MyBatis 行为的信息,这些信息通常只配置在一个文件中,并不轻易改动。另外,与 Spring MVC 框架整合后,MyBatis 的核心配置信息将配置到 Spring MVC 的配置文件或配置类中。因此,在实际开发中,很少编写或修改 MyBatis 的核心配置文件。本节仅了解 MyBatis 核心配置文件的主要元素。

MyBatis 的核心配置文件模板代码如下:

```xml
<?xml version="1.0" encoding="UTF-8" ?>
<!DOCTYPE configuration
PUBLIC "-//mybatis.org//DTD Config 3.0//EN"
"http://mybatis.org/dtd/mybatis-3-config.dtd">
<configuration>
    <properties/><!--属性 -->
    <settings><!--设置 -->
        <setting name="" value=""/>
    </settings>
    <typeAliases/><!--类型命名(别名) -->
    <typeHandlers/><!--类型处理器 -->
    <objectFactory type=""/><!--对象工厂 -->
    <plugins><!--插件 -->
        <plugin interceptor=""></plugin>
    </plugins>
    <environments default=""><!--配置环境 -->
        <environment id=""><!--环境变量 -->
            <transactionManager type=""/><!--事务管理器 -->
            <dataSource type=""/><!--数据源 -->
        </environment>
    </environments>
    <databaseIdProvider type=""/><!--数据库厂商标识 -->
    <mappers><!--映射器,告诉 MyBatis 到哪里去找映射文件-->
        <mapper resource="com/mybatis/UserMapper.xml"/>
    </mappers>
</configuration>
```

MyBatis 核心配置文件中的元素配置顺序不能颠倒,否则在 MyBatis 启动阶段就将发生异常。

13.2 映射器概述

映射器是 MyBatis 最复杂且最重要的组件,由一个接口加一个 XML 文件(SQL 映射文件)组成(见 12.5.3 节)。MyBatis 的映射器也可以使用注解完成,但在实际应用中使用不多,原因主要来自这几个方面:其一,面对复杂的 SQL 会显得无力;其二,注解的可读性较差;其三,注解丢失了 XML 上下文相互引用的功能。因此,推荐使用 XML 文件开发 MySQL 的映射器。

SQL 映射文件的常用配置元素如表 13.1 所示。

表 13.1 SQL 映射文件的常用配置元素

元素名称	描 述	备 注
select	查询语句,最常用、最复杂的元素之一	可以自定义参数,返回结果集等
insert	插入语句	执行后返回一个整数,代表插入的行数
update	更新语句	执行后返回一个整数,代表更新的行数
delete	删除语句	执行后返回一个整数,代表删除的行数
sql	定义一部分 SQL,在多个位置被引用	例如,一张表列名、一次定义,可以在多个 SQL 语句中使用
resultMap	用来描述从数据库结果集中来加载对象,是最复杂、最强大的元素	提供映射规则

13.3 <select> 元素

在 SQL 映射文件中,<select>元素用于映射 SQL 的 select 语句,其示例代码如下:

```
<!--根据 uid 查询一个用户信息 -->
<select id = "selectUserById" parameterType = "Integer" resultType = "com.po.MyUser">
    select * from user where uid = #{uid}
</select>
```

在上述示例代码中,id 的值是唯一标识符(对应 DAO 接口的某个方法),它接收一个 Integer 类型的参数,返回一个 MyUser 类型的对象,结果集自动映射到 MyUser 的属性。但需要注意的是,MyUser 的属性名称一定与查询结果的列名相同。

除上述示例代码中几个属性外,<select>元素还有一些常用的属性,如表 13.2 所示。

表 13.2 <select>元素的常用属性

属性名称	描 述
id	它和 Mapper 的命名空间组合起来使用(对应 DAO 接口的某个方法),是唯一标识符,供 MyBatis 调用
parameterType	表示传入 SQL 语句的参数类型的全限定名或别名。是个可选属性,MyBatis 能推断出具体传入语句的参数
resultType	SQL 语句执行后返回的类型(全限定名或者别名)。如果是集合类型,返回的是集合元素的类型。返回时可以使用 resultType 或 resultMap 之一

续表

属性名称	描　　述
resultMap	它是映射集的引用,与<resultMap>元素一起使用。返回时可以使用resultType或resultMap之一
flushCache	它的作用是在调用SQL语句后,是否要求MyBatis清空之前查询本地缓存和二级缓存,默认值为false。如果设置为true,则任何时候只要SQL语句被调用,都将清空本地缓存和二级缓存
useCache	启动二级缓存的开关。默认值为true,表示将查询结果存入二级缓存中
timeout	用于设置超时参数,单位是秒。超时将抛出异常
fetchSize	获取记录的总条数设定
statementType	告诉MyBatis使用哪个JDBC的Statement工作,取值为STATEMENT(Statement)、PREPARED(PreparedStatement)、CALLABLE(CallableStatement),默认值为PREPARED
resultSetType	这是针对JDBC的ResultSet接口而言,其值可设置为FORWARD_ONLY(只允许向前访问)、SCROLL_SENSITIVE(双向滚动,但不及时更新)、SCROLL_INSENSITIVE(双向滚动,及时更新)

13.3.1　使用Map接口传递参数

在实际开发中,查询SQL语句经常需要多个参数,比如多条件查询。传递多个参数时,<select>元素的parameterType属性值的类型是什么呢？在MyBatis中,允许Map接口通过键值对传递多个参数。

假设数据操作接口中有个实现查询陈姓男性用户信息功能的方法：

```
public List<MyUser>selectAllUser(Map<String, Object>param);
```

此时,传递给MyBatis映射器的是一个Map对象,使用该Map对象在SQL中设置对应的参数,对应SQL映射文件代码如下：

```
<!--查询陈姓男性用户信息 -->
<select id="selectAllUser" resultType="MyUser" parameterType="map">
    select * from user
    where uname like concat('%',#{u_name},'%')
    and usex =#{u_sex}
</select>
```

上述SQL映射文件中参数名u_name和u_sex是Map中的key。

【例13-1】　测试Map接口传递参数。

具体实现步骤如下。

1. 使用STS创建Maven项目并添加相关依赖

使用STS创建一个名为ch13_1的Maven Project,并通过pom.xml文件添加项目所依赖的 JAR 包(包括 MyBatis、Spring MVC、Spring JDBC、MySQL 连接器、MyBatis 与 Spring 桥接器、Log4j、DBCP 以及 JSTL 等依赖)。ch13_1 的 pom.xml 文件内容如下:

```xml
<project xmlns="http://maven.apache.org/POM/4.0.0"
    xmlns:xsi="http://www.w3.org/2001/XMLSchema-instance"
    xsi:schemaLocation="http://maven.apache.org/POM/4.0.0
https://maven.apache.org/xsd/maven-4.0.0.xsd">
    <modelVersion>4.0.0</modelVersion>
    <groupId>com.cn.chenheng</groupId>
    <artifactId>ch13_1</artifactId>
    <version>0.0.1-SNAPSHOT</version>
    <packaging>war</packaging>
    <properties>
        <!--spring 版本号 -->
        <spring.version>5.2.3.RELEASE</spring.version>
    </properties>
    <dependencies>
        <dependency>
            <groupId>org.springframework</groupId>
            <artifactId>spring-webmvc</artifactId>
            <version>${spring.version}</version>
        </dependency>
        <dependency>
            <groupId>org.springframework</groupId>
            <artifactId>spring-jdbc</artifactId>
            <version>${spring.version}</version>
        </dependency>
        <dependency>
            <groupId>mysql</groupId>
            <artifactId>mysql-connector-java</artifactId>
            <version>5.1.45</version>
        </dependency>
        <dependency>
            <groupId>org.apache.commons</groupId>
            <artifactId>commons-dbcp2</artifactId>
            <version>2.7.0</version>
        </dependency>
        <dependency>
            <groupId>org.mybatis</groupId>
            <artifactId>mybatis</artifactId>
            <version>3.5.4</version>
        </dependency>
        <dependency>
            <groupId>org.mybatis</groupId>
```

```xml
            <artifactId>mybatis-spring</artifactId>
            <version>2.0.3</version>
        </dependency>
        <dependency>
            <groupId>log4j</groupId>
            <artifactId>log4j</artifactId>
            <version>1.2.17</version>
        </dependency>
        <dependency>
            <groupId>javax.servlet</groupId>
            <artifactId>jstl</artifactId>
            <version>1.2</version>
        </dependency>
    </dependencies>
</project>
```

2. 创建数据库连接信息属性文件及Log4j的日志配置文件

在应用 ch13_1 的 src/main/resources 目录下创建数据库连接信息属性文件 jdbc.properties 文件,其内容与 12.5.3 节相同,不再赘述。

在应用 ch13_1 的 src/main/resources 目录下创建 Log4j 的日志配置文件 log4j.properties 文件,其内容与 12.4.2 节相同,不再赘述。

3. 创建持久化类

在应用 ch13_1 的 src/main/java 目录下创建一个名为 com.mybatis.po 的包,并在该包中创建持久化类 MyUser。该类与 12.4.3 节相同,不再赘述。

4. 创建SQL映射文件

在应用 ch13_1 的 src/main/resources 目录下创建一个名为 com.mybatis.mapper 的包,并在该包中创建 SQL 映射文件 UserMapper.xml。在该文件中编写 SQL 语句实现陈姓男性用户信息的查询,具体代码如下:

```xml
<?xml version="1.0" encoding="UTF-8" ?>
<!DOCTYPE mapper
PUBLIC "-//mybatis.org//DTD Mapper 3.0//EN"
"http://mybatis.org/dtd/mybatis-3-mapper.dtd">
<mapper namespace="com.mybatis.mapper.UserMapper">
    <!--查询陈姓男性用户信息 -->
    <select id="selectAllUser" resultType="MyUser" parameterType="map">
        select * from user
        where uname like concat('%',#{u_name},'%')
        and usex =#{u_sex}
    </select>
```

```
</mapper>
```

5. 创建 MyBatis 的核心配置文件

在应用 ch13_1 的 src/main/resources 目录下创建 MyBatis 的核心配置文件 mybatis-config.xml。在该文件中指定类型别名和 SQL 映射文件的位置,而数据源的配置在 Spring MVC 的 Java 配置类中完成。mybatis-config.xml 的内容如下:

```xml
<?xml version="1.0" encoding="UTF-8" ?>
<!DOCTYPE configuration
PUBLIC "-//mybatis.org//DTD Config 3.0//EN"
"http://mybatis.org/dtd/mybatis-3-config.dtd">
<configuration>
    <settings>
        <!--输出日志配置 -->
        <setting name="logImpl" value="LOG4J" />
    </settings>
    <typeAliases>
        <!--指定类型别名 -->
        <typeAlias type="com.mybatis.po.MyUser" alias="MyUser"/>
    </typeAliases>
    <mappers>
        <!--映射文件的位置 -->
        <mapper resource="com/mybatis/mapper/UserMapper.xml"/>
    </mappers>
</configuration>
```

6. 创建 Mapper 接口

在应用 ch13_1 的 src/main/java 目录下创建一个名为 com.mybatis.mapper 的包,并在该包中创建 Mapper 接口 UserMapper。该接口中的方法与 com.mybatis.mapper 中的 SQL 映射文件 UserMapper.xml 的 id 一致。UserMapper 接口代码如下:

```java
package com.mybatis.mapper;
import java.util.List;
import java.util.Map;
import org.springframework.stereotype.Repository;
import com.mybatis.po.MyUser;
@Repository
public interface UserMapper {
    public List<MyUser> selectAllUser(Map<String, Object> param);
```

}

7. 创建控制类

在应用 ch13_1 的 src/main/java 目录下创建一个名为 controller 的包,并在该包中创建控制器类 TestController。在该控制器类中调用 Mapper 接口中的方法操作数据库,具体代码如下:

```java
package controller;
import java.util.HashMap;
import java.util.List;
import java.util.Map;
import org.springframework.beans.factory.annotation.Autowired;
import org.springframework.stereotype.Controller;
import org.springframework.ui.Model;
import org.springframework.web.bind.annotation.RequestMapping;
import com.mybatis.mapper.UserMapper;
import com.mybatis.po.MyUser;
@Controller
public class TestController {
    @Autowired
    private UserMapper userMapper;
    @RequestMapping("/selectByUnameAndUsex")
    public String test(Model model) {
        //查询所有陈姓男性用户
        Map<String, Object>map =new HashMap<>();
        map.put("u_name", "陈");
        map.put("u_sex", "男");
        List<MyUser>unameAndUsexList =userMapper.selectAllUser(map);
        model.addAttribute("unameAndUsexList", unameAndUsexList);
        return "showUnameAndUsexUser";
    }
}
```

8. 创建查询结果显示页面

在应用 ch13_1 的 src/main/webapp/WEB-INF/ 目录下创建一个名为 jsp 的文件夹,并在该文件夹中创建 showUnameAndUsexUser.jsp 文件显示查询结果,具体代码如下:

```jsp
<%@page language="java" contentType="text/html; charset=UTF-8" pageEncoding="UTF-8"%>
<%@taglib prefix="c" uri="http://java.sun.com/jsp/jstl/core" %>
<%
String path =request.getContextPath();
String basePath = request.getScheme() +"://"+ request.getServerName() +":"+
```

```
request.getServerPort()+path+"/";
%>
<!DOCTYPE html>
<html>
<head>
<base href="<%=basePath%>">
<meta charset="UTF-8">
<title>Insert title here</title>
<link rel="stylesheet" href="static/css/bootstrap.min.css" />
</head>
<body>
    <div class="container">
        <div class="panel panel-primary">
            <div class="panel-heading">
                <h3 class="panel-title">陈姓男性用户列表</h3>
            </div>
            <div class="panel-body">
                <div class="table table-responsive">
                    <table class="table table-bordered table-hover">
                        <tbody class="text-center">
                            <tr>
                                <th>用户 ID</th>
                                <th>姓名</th>
                                <th>性别</th>
                            </tr>
                            <c:forEach items="${unameAndUsexList}" var="user">
                                <tr>
                                    <td>${user.uid}</td>
                                    <td>${user.uname}</td>
                                    <td>${user.usex}</td>
                                </tr>
                            </c:forEach>
                        </tbody>
                    </table>
                </div>
            </div>
        </div>
    </div>
</body>
</html>
```

9. 创建 Web、Spring MVC 和 Spring JDBC 的配置类

在应用 ch13_1 的 src/main/java 目录下创建一个名为 config 的包，并在该包中创建

Web 的配置类 WebConfig、Spring MVC 的配置类 SpringMVCConfig 以及 Spring JDBC 的配置类 SpringJDBCConfig。

在 SpringMVCConfig 类中配置视图解析器和静态资源（不需要 DispatcherServlet 转发的请求），具体代码如下：

```
package config;
import org.springframework.context.annotation.Bean;
import org.springframework.context.annotation.ComponentScan;
import org.springframework.context.annotation.Configuration;
import org.springframework.transaction.annotation.EnableTransactionManagement;
import org.springframework.web.servlet.config.annotation.EnableWebMvc;
import org.springframework.web.servlet.config.annotation.ResourceHandlerRegistry;
import org.springframework.web.servlet.config.annotation.WebMvcConfigurer;
import org.springframework.web.servlet.view.InternalResourceViewResolver;
@Configuration // 通过该注解来表明该类是一个 Spring 的配置，相当于一个 xml 文件
@EnableWebMvc // 开启 Spring MVC 的支持
@ComponentScan(basePackages = { "controller"}) // 扫描基本包
//配置多个配置文件  value={"classpath:jdbc.properties","xx","xxx"}
@EnableTransactionManagement // 开启声明式事务的支持
public class SpringMVCConfig implements WebMvcConfigurer{
    /**
     * 配置视图解析器
     */
    @Bean
    public InternalResourceViewResolver getViewResolver() {
        InternalResourceViewResolver viewResolver =new
        InternalResourceViewResolver();
        viewResolver.setPrefix("/WEB-INF/jsp/");
        viewResolver.setSuffix(".jsp");
        return viewResolver;
    }
    /**
     * 配置静态资源(不需要 DispatcherServlet 转发的请求)
     */
    @Override
    public void addResourceHandlers(ResourceHandlerRegistry registry) {
        registry.addResourceHandler("/static/**").addResourceLocations
        ("/static/");
    }
}
```

在 SpringJDBCConfig 配置类中配置视数据源，为数据源添加事务管理器以及配置 MyBatis 工厂，具体代码如下：

```
package config;
```

```java
import org.apache.commons.dbcp2.BasicDataSource;
import org.mybatis.spring.SqlSessionFactoryBean;
import org.mybatis.spring.annotation.MapperScan;
import org.springframework.beans.factory.annotation.Value;
import org.springframework.context.annotation.Bean;
import org.springframework.context.annotation.Configuration;
import org.springframework.context.annotation.PropertySource;
import org.springframework.core.io.ClassPathResource;
import org.springframework.core.io.Resource;
import org.springframework.jdbc.datasource.DataSourceTransactionManager;
import org.springframework.transaction.annotation.EnableTransactionManagement;
@Configuration // 通过该注解来表明该类是一个 Spring 的配置,相当于一个 xml 文件
@PropertySource(value = { "classpath:jdbc.properties" },
ignoreResourceNotFound = true)
//配置多个配置文件    value={"classpath:jdbc.properties","xx","xxx"}
@EnableTransactionManagement // 开启声明式事务的支持
@MapperScan(basePackages = { "com.mybatis.mapper" }) // 配置扫描 MyBatis 接口的包
路径
public class SpringJDBCConfig {
    @Value("${db.url}") // 注入属性文件 jdbc.properties 中的 jdbc.url
    private String jdbcUrl;
    @Value("${db.driverClassName}")
    private String jdbcDriverClassName;
    @Value("${db.username}")
    private String jdbcUsername;
    @Value("${db.password}")
    private String jdbcPassword;
    @Value("${db.maxTotal}")
    private int maxTotal;
    @Value("${db.maxIdle}")
    private int maxIdle;
    @Value("${db.initialSize}")
    private int initialSize;
    /**
     * 配置数据源
     */
    @Bean
    public BasicDataSource dataSource() {
        BasicDataSource myDataSource = new BasicDataSource();
        // 数据库驱动
        myDataSource.setDriverClassName(jdbcDriverClassName);
        // 相应驱动的 jdbcUrl
        myDataSource.setUrl(jdbcUrl);
        // 数据库的用户名
```

```java
        myDataSource.setUsername(jdbcUsername);
        // 数据库的密码
        myDataSource.setPassword(jdbcUsername);
        // 最大连接数
        myDataSource.setMaxTotal(maxTotal);
        // 最大空闲连接数
        myDataSource.setMaxIdle(maxIdle);
        // 初始化连接数
        myDataSource.setInitialSize(initialSize);
        return myDataSource;
    }
    /**
     * 为数据源添加事务管理器
     */
    @Bean
    public DataSourceTransactionManager transactionManager() {
        DataSourceTransactionManager dt =new DataSourceTransactionManager();
        dt.setDataSource(dataSource());
        return dt;
    }
    /**
     * 配置MyBatis工厂,同时指定数据源,并与MyBatis完美整合
     */
    @Bean
    public SqlSessionFactoryBean getSqlSession() {
        SqlSessionFactoryBean sqlSessionFactory =new SqlSessionFactoryBean();
        sqlSessionFactory.setDataSource(dataSource());
        Resource r =new ClassPathResource("mybatis-config.xml");
        sqlSessionFactory.setConfigLocation(r);
        return sqlSessionFactory;
    }
}
```

在WebConfig配置类中注册SpringJDBCConfig、SpringMVCConfig以及Spring MVC的DispatcherServlet,具体代码如下:

```java
package config;
import javax.servlet.ServletContext;
import javax.servlet.ServletException;
import javax.servlet.ServletRegistration.Dynamic;
import org.springframework.web.WebApplicationInitializer;
import org.springframework.web.context.support.AnnotationConfigWebApplicationContext;
import org.springframework.web.servlet.DispatcherServlet;
public class WebConfig implements WebApplicationInitializer{
    @Override
```

```
public void onStartup(ServletContext arg0) throws ServletException {
    AnnotationConfigWebApplicationContext ctx
    =new AnnotationConfigWebApplicationContext();
    ctx.register(SpringJDBCConfig.class);//注册配置类SpringJDBCConfig
    ctx.register(SpringMVCConfig.class);//注册配置类SpringMVCConfig
    ctx.setServletContext(arg0);//和当前ServletContext关联
    /**
     * 注册Spring MVC的DispatcherServlet
     */
    Dynamic servlet = arg0.addServlet("dispatcher", new DispatcherServlet
    (ctx));
    servlet.addMapping("/");
    servlet.setLoadOnStartup(1);
}
}
```

10. 测试应用

发布应用 ch13_1 到 Web 服务器 Tomcat 后,通过地址 http://localhost:8080/ch13_1/selectByUnameAndUsex 测试应用。成功运行后的效果如图 13.1 所示。

图 13.1　查询所有陈姓男性用户

Map 是一个键值对应的集合,使用者需要通过阅读它的键才能了解其作用。另外,使用 Map 不能限定其传递的数据类型,所以业务性不强,可读性差。如果 SQL 语句很复杂,参数很多,使用 Map 就很不方便。幸运的是,MyBatis 还提供使用 Java Bean 传递参数。

13.3.2　使用 Java Bean 传递参数

在 MyBatis 中,当需要将多个参数传递给映射器时,可以将它们封装在一个 Java Bean 中。下面通过具体实例讲解如何使用 Java Bean 传递参数。

【例 13-2】　使用 Java Bean 传递参数。

该实例在【例 13-1】的基础上实现,相同的实现不再赘述。其他的具体实现如下。

1. 添加 Mapper 接口方法

在 UserMapper 接口中添加数据操作接口方法 selectAllUserByJavaBean(),在该方法中使用 MyUser 类的对象封装参数信息。接口方法 selectAllUserByJavaBean()的定义如下:

```
public List<MyUser>selectAllUserByJavaBean(MyUser user);
```

2. 添加 SQL 映射

在 SQL 映射文件 UserMapper.xml 中添加接口方法 selectAllUserByJavaBean()对应的 SQL 映射,具体代码如下:

```xml
<!--通过 Java Bean 传递参数查询陈姓男性用户信息,#{uname}的 uname 为参数 MyUser 的属性-->
<select id="selectAllUserByJavaBean" resultType="MyUser" parameterType="MyUser">
    select * from user
    where uname like concat('%',#{uname},'%')
    and usex =#{usex}
</select>
```

3. 添加请求处理方法

在控制器类 UserController 中添加请求处理方法 selectAllUserByJavaBean(),具体代码如下:

```java
/**
 * 通过 Java Bean 传递参数
 */
@RequestMapping("/selectAllUserByJavaBean")
public String selectAllUserByJavaBean(Model model) {
    //通过 MyUser 封装参数,查询所有陈姓男性用户
    MyUser mu =new MyUser();
    mu.setUname("陈");
    mu.setUsex("男");
    List<MyUser>unameAndUsexList =userMapper.selectAllUserByJavaBean(mu);
    model.addAttribute("unameAndUsexList", unameAndUsexList);
    return "showUnameAndUsexUser";
}
```

4. 测试应用

重启 Web 服务器 Tomcat,通过地址 http://localhost:8080/ch13_1/selectAllUserByJavaBean

13.3.3 使用@Param 注解传递参数

不管是 Map 传参,还是 Java Bean 传参,都是将多个参数封装在一个对象中,实际传递的还是一个参数。而使用@Param 注解可以将多个参数依次传递给 MyBatis 映射器。示例代码如下:

```
public List<MyUser> selectByParam(@Param("puname") String uname, @Param("pusex") String usex);
```

在上述示例代码中,puname 和 pusex 是传递给 MyBatis 映射器的参数名。下面通过实例讲解如何使用@Param 注解传递参数。

【例 13-3】 使用@Param 注解传递参数。

该实例在【例 13-1】的基础上实现,相同的实现不再赘述。其他的具体实现如下。

1. 添加 Mapper 接口方法

在 UserMapper 接口中添加数据操作接口方法 selectAllUserByParam(),在该方法中使用@Param 注解传递两个参数。接口方法 selectAllUserByParam()的定义如下:

```
public List<MyUser> selectAllUserByParam(@Param("puname") String uname, @Param("pusex") String usex);
```

2. 添加 SQL 映射

在 SQL 映射文件 UserMapper.xml 中添加接口方法 selectAllUserByParam()对应的 SQL 映射,具体代码如下:

```xml
<!--通过@Param 注解传递参数查询陈姓男性用户信息,这里不需要定义参数类型-->
<select id="selectAllUserByParam" resultType="MyUser">
    select * from user
    where uname like concat('%',#{puname},'%')
    and usex = #{pusex}
</select>
```

3. 添加请求处理方法

在控制器类 UserController 中添加请求处理方法 selectAllUserByParam(),具体代码如下:

```java
/**
 *通过@Param 注解传递参数
 */
@RequestMapping("/selectAllUserByParam")
public String selectAllUserByParam(Model model) {
```

```
//通过@Param注解传递参数,查询所有陈姓男性用户
    List<MyUser> unameAndUsexList = userMapper.selectAllUserByParam("陈",
"男");
    model.addAttribute("unameAndUsexList", unameAndUsexList);
    return "showUnameAndUsexUser";
}
```

4. 测试应用

重启 Web 服务器 Tomcat,通过地址 http://localhost:8080/ch13_1/selectAllUserByParam 测试应用。

在实际应用中是选择 Map、Java Bean,还是选择@Param 传递多个参数,应根据实际情况决定。如果参数较少,建议选择@Param;如果参数较多,建议选择 Java Bean。

13.3.4 ＜resultMap＞元素

＜resultMap＞元素表示结果映射集,是 MyBatis 中最重要也是最强大的元素。主要用来定义映射规则、级联的更新以及定义类型转化器等。＜resultMap＞元素包含了一些子元素,结构如下:

```
<resultMap type="" id="">
    <constructor><!--类在实例化时,用来注入结果到构造方法 -->
        <idArg/><!--ID参数,结果为 ID -->
        <arg/><!--注入到构造方法的一个普通结果 -->
    </constructor>
    <id/><!--用于表示哪个列是主键 -->
    <result/><!--注入到字段或 POJO 属性的普通结果 -->
    <association property=""/><!--用于一对一关联 -->
    <collection property=""/><!--用于一对多、多对多关联 -->
    <discriminator javaType=""><!--使用结果值来决定使用哪个结果映射 -->
        <case value=""/><!--基于某些值的结果映射 -->
    </discriminator>
</resultMap>
```

＜resultMap＞元素的 type 属性表示需要的 POJO,id 属性是 resultMap 的唯一标识。子元素＜constructor＞用于配置构造方法(当 POJO 未定义无参数的构造方法时使用)。子元素＜id＞用于表示哪个列是主键。子元素＜result＞用于表示 POJO 和数据表普通列的映射关系。子元素＜association＞＜collection＞和＜discriminator＞是用在级联的情况下。级联的问题比较复杂,将在 13.7 节级联那里学习。

一条查询 SQL 语句执行后,将返回结果,而结果可以使用 Map 存储,也可以使用 POJO(Java Bean)存储。

13.3.5 使用 POJO 存储结果集

13.3.1 小节至 13.3.3 小节都是直接使用 Java Bean(MyUser)存储的

结果集,这是因为 MyUser 的属性名与查询结果集的列名相同。如果查询结果集的列名与 Java Bean 的属性名不同,那么可以结合＜resultMap＞元素,将 Java Bean 的属性与查询结果集的列名一一对应。

下面通过一个实例讲解如何使用＜resultMap＞元素将 Java Bean 的属性与查询结果集的列名一一对应。

【例 13-4】 使用＜resultMap＞元素将 Java Bean 的属性与查询结果集的列名一一对应。该实例在【例 13-1】的基础上实现,相同的实现不再赘述。其他的具体实现如下。

1. 创建 POJO 类

在应用 ch13_1 的 com.mybatis.po 包中创建一个名为 MapUser 的 POJO(Plain Ordinary Java Object,普通的 Java 类)类,具体代码如下:

```
package com.mybatis.po;
public class MapUser {
    private Integer m_uid;
    private String m_uname;
    private String m_usex;
    //此处省略 setter 和 getter 方法
    @Override
    public String toString() {
        return "User [uid=" +m_uid +",uname=" +m_uname +",usex=" +m_usex +"]";
    }
}
```

2. 添加 Mapper 接口方法

在 UserMapper 接口中添加数据操作接口方法 selectAllUserPOJO(),该方法的返回值类型是 List＜MapUser＞。接口方法 selectAllUserPOJO()的定义如下:

```
public List<MapUser>selectAllUserPOJO();
```

3. 添加 SQL 映射

打开 SQL 映射文件 UserMapper.xml,首先,在其中使用＜resultMap＞元素,将 MapUser 类的属性与查询结果列名一一对应。然后,添加接口方法 selectAllUserPOJO() 对应的 SQL 映射。具体代码如下:

```xml
<!--使用自定义结果集类型 -->
<resultMap type="com.mybatis.po.MapUser" id="myResult">
    <!--property 是 com.mybatis.po.MapUser 类中的属性-->
    <!--column 是查询结果的列名,可以来自不同的表 -->
    <id property="m_uid" column="uid"/>
    <result property="m_uname" column="uname"/>
```

```xml
        <result property="m_usex" column="usex"/>
</resultMap>
<!--使用自定义结果集类型查询所有用户 -->
<select id="selectAllUserPOJO" resultMap="myResult">
    select * from user
</select>
```

4. 添加请求处理方法

在控制器类 UserController 中添加请求处理方法 selectAllUserPOJO()，具体代码如下：

```java
/**
 * 使用自定义结果集类型查询所有用户
 */
@RequestMapping("/selectAllUserPOJO")
public String selectAllUserPOJO(Model model) {
    //通过@Param注解传递参数，查询所有陈姓男性用户
    List<MapUser>unameAndUsexList =userMapper.selectAllUserPOJO();
    model.addAttribute("unameAndUsexList", unameAndUsexList);
    return "showUnameAndUsexUserPOJO";
}
```

5. 创建显示查询结果的页面

在应用 ch13_1 的 src/main/webapp/WEB-INF/jsp 目录下创建 showUnameAndUsexUserPOJO.jsp 文件，显示查询结果，具体代码如下：

```jsp
<%@page language="java" contentType="text/html; charset=UTF-8" pageEncoding="UTF-8"%>
<%@taglib prefix="c" uri="http://java.sun.com/jsp/jstl/core" %>
<%
String path =request.getContextPath();
String basePath = request.getScheme() +"://"+ request.getServerName() +":"+ request.getServerPort()+path+"/";
%>
<!DOCTYPE html>
<html>
<head>
<base href="<%=basePath%>">
<meta charset="UTF-8">
<title>Insert title here</title>
<link rel="stylesheet" href="static/css/bootstrap.min.css" />
</head>
<body>
```

```html
            <div class="container">
                <div class="panel panel-primary">
                    <div class="panel-heading">
                        <h3 class="panel-title">陈姓男性用户列表</h3>
                    </div>
                    <div class="panel-body">
                        <div class="table table-responsive">
                            <table class="table table-bordered table-hover">
                                <tbody class="text-center">
                                    <tr>
                                        <th>用户ID</th>
                                        <th>姓名</th>
                                        <th>性别</th>
                                    </tr>
                                    <c:forEach items="${unameAndUsexList}" var="user">
                                    <tr>
                                        <td>${user.m_uid}</td>
                                        <td>${user.m_uname}</td>
                                        <td>${user.m_usex}</td>
                                    </tr>
                                    </c:forEach>
                                </tbody>
                            </table>
                        </div>
                    </div>
                </div>
            </div>
</body>
</html>
```

6. 测试应用

重启 Web 服务器 Tomcat，通过地址 http://localhost:8080/ch13_1/selectAllUserPOJO 测试应用。成功运行后的效果如图 13.2 所示。

13.3.6　使用 Map 存储结果集

在 MyBatis 中，任何查询结果都可以使用 Map 存储。下面通过一个实例讲解如何使用 Map 存储查询结果。

【例 13-5】　使用 Map 存储查询结果。该实例在【例 13-1】的基础上实现，相同的实现不再赘述。其他的具体实现如下。

1. 添加 Mapper 接口方法

在 UserMapper 接口中添加数据操作接口方法 selectAllUserMap()，该方法的返回

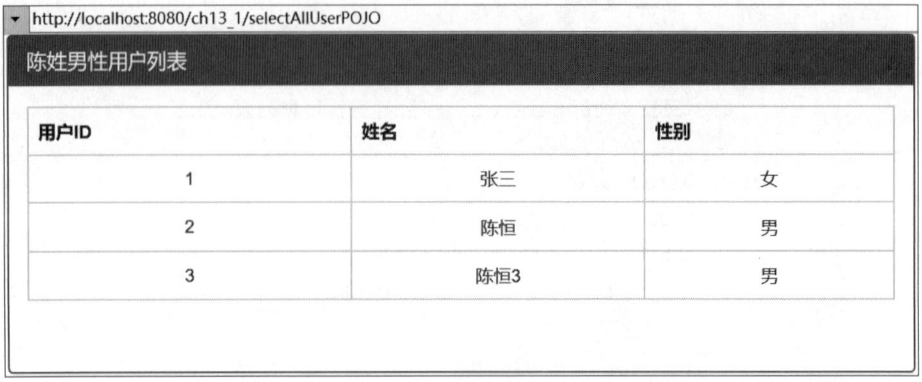
http://localhost:8080/ch13_1/selectAllUserPOJO

陈姓男性用户列表

用户ID	姓名	性别
1	张三	女
2	陈恒	男
3	陈恒3	男

图 13.2　查询结果

值类型是 List＜Map＜String, Object＞＞。接口方法 selectAllUserMap()的定义如下：

```
public List<Map<String, Object>> selectAllUserMap();
```

2. 添加 SQL 映射

在 SQL 映射文件 UserMapper.xml 中添加接口方法 selectAllUserMap()对应的 SQL 映射，具体代码如下：

```
<!--使用 Map 存储查询结果,查询结果的列名作为 Map 的 key,列值为 Map 的 value -->
<select id="selectAllUserMap" resultType="map">
    select * from user
</select>
```

3. 添加请求处理方法

在控制器类 UserController 中添加请求处理方法 selectAllUserMap()，具体代码如下：

```
/**
 * 使用 Map 存储查询结果
 */
@RequestMapping("/selectAllUserMap")
public String selectAllUserMap(Model model) {
    //使用 Map 存储查询结果
    List<Map<String, Object>> unameAndUsexList = userMapper.selectAllUserMap();
    model.addAttribute("unameAndUsexList", unameAndUsexList);
    //在 showUnameAndUsexUser.jsp 页面中遍历时,属性名与查询结果的列名(Map 的 key)
      相同
    return "showUnameAndUsexUser";
}
```

4. 测试应用

重启 Web 服务器 Tomcat，通过地址 http://localhost:8080/ch13_1/selectAllUserMap 测试应用。

13.3.7 实践环节

查询用户 id 大于 3 的张姓女性用户信息。要求使用 @Param 注解传递参数，使用 Map 存储结果集。

13.4 <insert> 元素

<insert>元素用于映射插入语句，MyBatis 执行完一条插入语句后，将返回一个整数，表示其影响的行数。它的属性与<select>元素的属性大部分相同，在本节讲解它的几个特有属性。

keyProperty：将插入或更新操作时的返回值赋值给 PO（Persistant Object）类的某个属性，通常会设置为主键对应的属性。如果是联合主键，可以在多个值之间用逗号隔开。

keyColumn：设置第几列是主键，当主键列不是表中的第一列时，需要设置。如果是联合主键时，可以在多个值之间用逗号隔开。

useGeneratedKeys：该属性将使 MyBatis 使用 JDBC 的 getGeneratedKeys() 方法获取由数据库内部生产的主键，如 MySQL、SQL Server 等自动递增的字段，默认值为 false。

13.4.1 主键（自动递增）回填

MySQL、SQL Server 等数据库的表格可以采用自动递增的字段作为主键。有时可能需要使用这个刚刚产生的主键，用以关联其他业务。因为本书采用的数据库是 MySQL 数据库，所以可以直接使用 ch13_1 应用讲解自动递增主键回填的使用方法。

【例 13-6】 自动递增主键回填的使用方法。

该实例在【例 13-1】的基础上实现，相同的实现不再赘述。其他的具体实现如下。

1. 添加 Mapper 接口方法

在 UserMapper 接口中添加数据操作接口方法 addUser()，该方法的返回值类型是 int。接口方法 addUser() 的定义如下：

```
public int addUser(MyUser mu);
```

2. 添加 SQL 映射

在 SQL 映射文件 UserMapper.xml 中添加接口方法 addUser() 对应的 SQL 映射，具体代码如下：

```xml
<!--添加一个用户,成功后将主键值回填给 uid(po类的属性)-->
<insert id="addUser" parameterType="MyUser" keyProperty="uid" useGeneratedKeys="true">
    insert into user (uname,usex) values(#{uname},#{usex})
</insert>
```

3. 添加请求处理方法

在控制器类 UserController 中添加请求处理方法 addUser(),具体代码如下:

```java
/**
 * 添加一个用户,成功后将主键值回填给 uid(MyUser 类的属性)
 */
@RequestMapping("/addUser")
public String addUser(Model model) {
    //添加一个用户
    MyUser addmu = new MyUser();
    addmu.setUname("陈恒主键回填");
    addmu.setUsex("男");
    userMapper.addUser(addmu);
    model.addAttribute("addmu", addmu);
    return "showAddUser";
}
```

4. 创建显示被添加的用户信息页面

在应用 ch13_1 的 src/main/webapp/WEB-INF/jsp 目录下创建 showAddUser.jsp 文件显示添加的用户信息,具体代码如下:

```jsp
<%@page language="java" contentType="text/html; charset=UTF-8" pageEncoding="UTF-8"%>
<%
String path = request.getContextPath();
String basePath = request.getScheme() + "://" + request.getServerName() + ":" + request.getServerPort()+path+"/";
%>
<!DOCTYPE html>
<html>
<head>
<base href="<%=basePath%>">
<meta charset="UTF-8">
<title>Insert title here</title>
<link rel="stylesheet" href="static/css/bootstrap.min.css" />
</head>
<body>
```

```html
        <div class="container">
            <div class="panel panel-primary">
                <div class="panel-heading">
                    <h3 class="panel-title">添加的用户信息</h3>
                </div>
                <div class="panel-body">
                    <div class="table table-responsive">
                        <table class="table table-bordered table-hover">
                            <tbody class="text-center">
                                <tr>
                                    <th>用户 ID(回填的主键)</th>
                                    <th>姓名</th>
                                    <th>性别</th>
                                </tr>
                                <tr>
                                    <td>${addmu.uid}</td>
                                    <td>${addmu.uname}</td>
                                    <td>${addmu.usex}</td>
                                </tr>
                            </tbody>
                        </table>
                    </div>
                </div>
            </div>
        </div>
</div>
</body>
</html>
```

5．测试应用

重启 Web 服务器 Tomcat，通过地址 http://localhost:8080/ch13_1/addUser 测试应用。运行结果如图 13.3 所示。

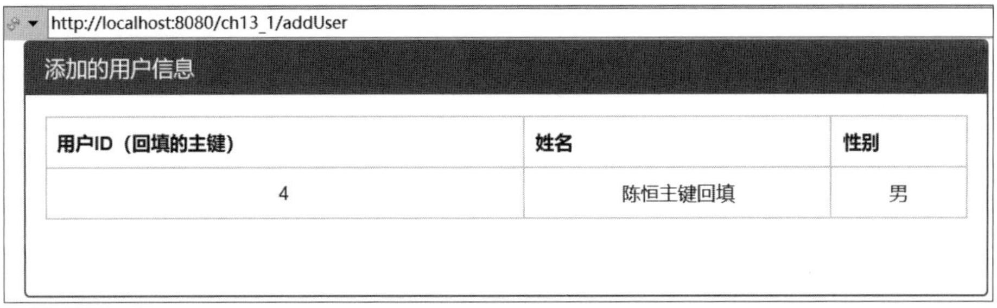

图 13.3　添加的用户信息

13.4.2 自定义主键

如果实际工程中使用的数据库不支持主键自动递增（如 Oracle），或者取消了主键自动递增的规则，可以使用 MyBatis 的<selectKey>元素来自定义生成主键。具体配置示例代码如下：

```
<insert id="insertUser" parameterType="MyUser">
    <!--先使用 selectKey 元素定义主键，然后再定义 SQL 语句 -->
    <selectKey keyProperty="uid" resultType="Integer" order="BEFORE">
        select decode(max(uid), null, 1, max(uid)+1) as newUid from user
    </selectKey>
    insert into user (uid,uname,usex) values(#uid,#{uname},#{usex})
</insert>
```

执行上述示例代码时，<selectKey>元素首先被执行，该元素通过自定义的语句设置数据表的主键，然后执行插入语句。

<selectKey>元素的 keyProperty 属性指定了新生主键值返回给 PO 类（MyUser）的哪个属性。order 属性可以设置为 BEFORE 或 AFTER，BEFORE 表示先执行<selectKey>元素，然后执行插入语句；AFTER 表示先执行插入语句，再执行<selectKey>元素。

13.5 <update>与<delete>元素

<update>和<delete>元素比较简单，它们的属性和<insert>元素的属性基本一样，执行后也返回一个整数，表示影响数据库的记录行数。配置示例代码如下：

```
<!--修改一个用户 -->
<update id="updateUser" parameterType="MyUser">
    update user set uname=#{uname},usex=#{usex} where uid=#{uid}
</update>
<!--删除一个用户 -->
<delete id="deleteUser" parameterType="Integer">
    delete from user where uid=#{uid}
</delete>
```

13.6 <sql>元素

<sql>元素的作用是定义 SQL 语句的一部分（代码片段），方便后续的 SQL 语句引用它，比如反复使用的列名。在 MyBatis 中，只需使用<sql>元素编写一次，便能在其他元素中引用它。配置示例代码如下：

```
<sql id="comColumns">id,uname,usex</sql>
```

```
<select id="selectUser" resultType="MyUser">
    select <include refid="comColumns"/>from user
</select>
```

在上述代码中,使用<include>元素的 refid 属性引用了自定义的代码片段。

13.7 级联查询

级联关系是一个数据库实体的概念。有三种级联关系,分别是一对一级联、一对多级联以及多对多级联。级联的优点是获取关联数据十分方便,但是级联过多会增加数据库系统的复杂度,同时降低系统的性能。在实际开发中,要根据实际情况判断是否需要使用级联。更新和删除的级联关系很简单,数据库内在机制即可完成。本节仅讲述级联查询的相关实现。

如果表 A 中有一个外键引用了表 B 的主键,A 表就是子表,B 表就是父表。当查询表 A 的数据时,通过表 A 的外键,也将返回表 B 的相关记录,这就是级联查询。例如,查询一个人的信息时,同时根据外键(身份证号)也将返回他的身份证信息。

13.7.1 一对一级联查询

一对一级联关系在现实生活中是十分常见的。比如一个大学生只有一张一卡通,一张一卡通只属于一个学生。再比如人与身份证的关系也是一对一的级联关系。

MyBatis 如何处理一对一级联查询呢?在 MyBatis 中,通过<resultMap>元素的子元素<association>处理这种一对一级联关系。在<association>元素中通常使用以下属性。

- property:指定映射到实体类的对象属性。
- column:指定表中对应的字段(即查询返回的列名)。
- javaType:指定映射到实体对象属性的类型。
- select:指定引入嵌套查询的子 SQL 语句,该属性用于关联映射中的嵌套查询。

下面以个人与身份证之间的关系为例,讲解一对一级联查询的处理过程,读者只需参考该实例即可学会一对一级联查询的 MyBatis 实现。

【例 13-7】 一对一级联查询的 MyBatis 实现。

具体实现步骤如下。

1. 创建数据表

本实例需要在数据库 springtest 中创建两张数据表,一张是身份证表 idcard,一张是个人信息表 person。这两张表具有一对一的级联关系,它们的创建代码如下:

```
CREATE TABLE 'idcard' (
  'id' int(11) NOT NULL AUTO_INCREMENT,
  'code' varchar(18) COLLATE utf8_unicode_ci DEFAULT NULL,
```

```
    PRIMARY KEY ('id')
);
CREATE TABLE 'person' (
  'id' int(11) NOT NULL AUTO_INCREMENT,
  'name' varchar(20) COLLATE utf8_unicode_ci DEFAULT NULL,
  'age' int(11) DEFAULT NULL,
  'idcard_id' int(11) DEFAULT NULL,
  PRIMARY KEY ('id'),
  KEY 'idcard_id' ('idcard_id'),
  CONSTRAINT 'idcard_id' FOREIGN KEY ('idcard_id') REFERENCES 'idcard' ('id')
);
```

2. 使用 STS 创建 Maven 项目并添加相关依赖

使用 STS 创建一个名为 ch13_2 的 Maven Project，并通过 pom.xml 文件添加项目所依赖的 JAR 包（包括 MyBatis、Spring MVC、Spring JDBC、MySQL 连接器、MyBatis 与 Spring 桥接器、Log4j、DBCP 以及 JSTL 等依赖）。ch13_2 的 pom.xml 文件内容与应用 ch13_1 相同，不再赘述。

3. 创建数据库连接信息属性文件及 Log4j 的日志配置文件

在应用 ch13_2 的 src/main/resources 目录下创建数据库连接信息属性文件 jdbc.properties 文件，其内容与应用 ch13_1 相同，不再赘述。

在应用 ch13_2 的 src/main/resources 目录下创建 Log4j 的日志配置文件 log4j.properties 文件，其内容如下：

```
#Global logging configuration
log4j.rootLogger=ERROR, stdout
#MyBatis logging configuration...
log4j.logger.com.dao=DEBUG
#Console output...
log4j.appender.stdout=org.apache.log4j.ConsoleAppender
log4j.appender.stdout.layout=org.apache.log4j.PatternLayout
log4j.appender.stdout.layout.ConversionPattern=%5p [%t] -%m%n
```

4. 创建持久化类

在 ch13_2 应用的 src/main/java 目录下创建一个名为 com.po 的包，并在该包中创建数据表对应的持久化类 Idcard 和 Person。

Idcard 的代码如下：

```
package com.po;
/**
 * springtest 数据库中 idcard 表的持久化类
```

```java
 */
public class Idcard {
    private Integer id;
    private String code;
    //省略 setter 和 getter 方法
    /**
     * 方便测试,重写了 toString 方法
     */
    @Override
    public String toString() {
        return "Idcard [id=" +id +",code="+code +"]";
    }
}
```

Person 的代码如下:

```java
package com.po;
/**
 * springtest 数据库中 person 表的持久化类
 */
public class Person {
    private Integer id;
    private String name;
    private Integer age;
    //个人身份证关联
    private Idcard card;
    //省略 setter 和 getter 方法
    @Override
    public String toString() {
        return "Person [id=" +id +",name=" +name +",age=" +age +",card=" +card +"] ";
    }
}
```

5. 创建 SQL 映射文件

在应用 ch13_2 的 src/main/resources 目录下创建一个名为 com.mybatis.mapper 的包,并在该包中创建两张表对应的映射文件 IdCardMapper.xml 和 PersonMapper.xml。在 PersonMapper.xml 文件中以 3 种方式实现"根据个人 id 查询个人信息"的功能,详情见代码备注。

IdCardMapper.xml 的代码如下:

```xml
<?xml version="1.0" encoding="UTF-8" ?>
<!DOCTYPE mapper
PUBLIC "-//mybatis.org//DTD Mapper 3.0//EN"
"http://mybatis.org/dtd/mybatis-3-mapper.dtd">
```

```xml
<mapper namespace="com.dao.IdCardDao">
    <select id="selectCodeById" parameterType="Integer" resultType="Idcard">
        select * from idcard where id=#{id}
    </select>
</mapper>
```

PersonMapper.xml 的代码如下：

```xml
<?xml version="1.0" encoding="UTF-8" ?>
<!DOCTYPE mapper
PUBLIC "-//mybatis.org//DTD Mapper 3.0//EN"
"http://mybatis.org/dtd/mybatis-3-mapper.dtd">
<mapper namespace="com.dao.PersonDao">
    <!--一对一 根据 id 查询个人信息:级联查询的第一种方法(嵌套查询,执行两个 SQL 语句) -->
    <resultMap type="Person" id="cardAndPerson1">
        <id property="id" column="id"/>
        <result property="name" column="name"/>
        <result property="age" column="age"/>
        <!--一对一关联查询 -->
        <association property="card" column="idcard_id" javaType="Idcard"
            select="com.dao.IdCardDao.selectCodeById"/>
    </resultMap>
    <select id="selectPersonById1" parameterType="Integer" resultMap=
      "cardAndPerson1">
        select * from person where id=#{id}
    </select>
    <!--一对一 根据 id 查询个人信息:级联查询的第二种方法(嵌套结果,执行一个 SQL 语句) -->
    <resultMap type="Person" id="cardAndPerson2">
        <id property="id" column="id"/>
        <result property="name" column="name"/>
        <result property="age" column="age"/>
        <!--一对一关联查询 -->
        <association property="card" javaType="Idcard">
            <id property="id" column="idcard_id"/>
            <result property="code" column="code"/>
        </association>
    </resultMap>
    <select id="selectPersonById2" parameterType="Integer" resultMap=
      "cardAndPerson2">
        select p.*,ic.code
        from person p, idcard ic
        where p.idcard_id =ic.id and p.id=#{id}
    </select>
    <!--一对一 根据 id 查询个人信息:连接查询(使用 POJO 存储结果) -->
    <select id="selectPersonById3" parameterType="Integer" resultType=
```

```
            "SelectPersonById">
        select p.*,ic.code
        from person p, idcard ic
        where p.idcard_id =ic.id and p.id=#{id}
    </select>
</mapper>
```

6. 创建 POJO 类

在应用 ch13_2 的 src/main/java 目录下创建一个名为 com.pojo 的包,并在该包中创建 POJO 类 SelectPersonById(第 5 步使用的 POJO 类)。

SelectPersonById 的代码如下:

```
package com.pojo;
public class SelectPersonById {
    private Integer id;
    private String name;
    private Integer age;
    private String code;
    //省略 setter 和 getter 方法
    @Override
    public String toString() {
        return "Person [id=" +id +",name=" +name +",age="
        +age +",code=" +code +"]";
    }
}
```

7. 创建 MyBatis 的核心配置文件

在应用 ch13_2 的 src/main/resources 目录下创建 MyBatis 的核心配置文件 mybatis-config.xml。在该文件中打开延迟加载的开关,指定类型别名和 SQL 映射文件的位置。mybatis-config.xml 的内容如下:

```
<?xml version="1.0" encoding="UTF-8" ?>
<!DOCTYPE configuration
PUBLIC "-//mybatis.org//DTD Config 3.0//EN"
"http://mybatis.org/dtd/mybatis-3-config.dtd">
<configuration>
    <settings>
        <!--输出日志配置 -->
        <setting name="logImpl" value="LOG4J" />
        <!--打开延迟加载的开关 -->
        <setting name="lazyLoadingEnabled" value="true"/>
        <!--将积极加载改为按需加载 -->
        <setting name="aggressiveLazyLoading" value="false"/>
```

```xml
        </settings>
        <typeAliases>
            <!--指定类型别名 -->
            <typeAlias type="com.po.Idcard" alias="Idcard"/>
            <typeAlias type="com.po.Person" alias="Person"/>
            <typeAlias type="com.pojo.SelectPersonById" alias=
            "SelectPersonById"/>
        </typeAliases>
        <mappers>
            <!--映射文件的位置 -->
            <mapper resource="com/mybatis/mapper/IdCardMapper.xml"/>
            <mapper resource="com/mybatis/mapper/PersonMapper.xml"/>
        </mappers>
</configuration>
```

8. 创建 Mapper 接口

在 ch13_2 应用的 src/main/java 目录下创建一个名为 com.dao 的包,并在该包中创建第 5 步映射文件对应的数据操作接口 IdCardDao 和 PersonDao。

IdCardDao 的代码如下:

```java
package com.dao;
import org.springframework.stereotype.Repository;
import com.po.Idcard;
@Repository("idCardDao")
public interface IdCardDao {
    public Idcard selectCodeById(Integer i);
}
```

PersonDao 的代码如下:

```java
package com.dao;
import org.springframework.stereotype.Repository;
import com.po.Person;
import com.pojo.SelectPersonById;
@Repository("personDao")
public interface PersonDao {
    public Person selectPersonById1(Integer id);
    public Person selectPersonById2(Integer id);
    public SelectPersonById selectPersonById3(Integer id);
}
```

9. 创建 Web、Spring MVC 和 Spring JDBC 的配置类

在应用 ch13_2 的 src/main/java 目录下创建一个名为 config 的包，并在该包中创建 Web 的配置类 WebConfig、Spring MVC 的配置类 SpringMVCConfig 以及 Spring JDBC 的配置类 SpringJDBCConfig。此处 3 个配置类的代码与应用 ch13_1 的基本相同，不再赘述。

10. 创建控制器类

在应用 ch13_2 的 src/main/java 目录下创建一个名为 com.controller 的包，并在该包中创建 OneToOneController 控制器类，在该类中调用第 8 步的接口方法。具体代码如下：

```
package com.controller;
import org.springframework.beans.factory.annotation.Autowired;
import org.springframework.stereotype.Controller;
import org.springframework.web.bind.annotation.RequestMapping;
import com.dao.PersonDao;
import com.po.Person;
import com.pojo.SelectPersonById;
@Controller
public class OneToOneController{
    @Autowired
    private PersonDao personDao;
    @RequestMapping("/oneToOneTest")
    public String oneToOneTest() {
 System.out.println("级联查询的第一种方法(嵌套查询,执行两个SQL语句)");
        Person p1 =personDao.selectPersonById1(1);
        System.out.println(p1);
        System.out.println("======================");
        System.out.println("级联查询的第二种方法(嵌套结果,执行一个SQL语句)");
        Person p2 =personDao.selectPersonById2(1);
        System.out.println(p2);
        System.out.println("======================");
        System.out.println("连接查询(使用POJO存储结果)");
        SelectPersonById p3 =personDao.selectPersonById3(1);
        System.out.println(p3);
        return "test";
    }
}
```

11. 创建测试页面

在应用 ch13_2 的 src/main/webapp/WEB-INF/目录下创建 jsp 文件夹，并在该文件

夹中创建测试页面test.jsp文件显示"测试成功"信息，代码略。

12. 测试应用

发布应用到Web服务器Tomcat，通过地址http://localhost:8080/ch13_2/oneToOneTest测试应用。测试时，需事先为数据表手动添加数据。运行结果如图13.4所示。

图13.4 一对一级联查询结果

13.7.2 一对多级联查询

在13.7.1小节讲解了MyBatis如何处理一对一级联查询，那么MyBatis又如何处理一对多的级联查询呢？实际生活中有许多一对多的关系，例如一个用户可以有多个订单，而一个订单只属于一个用户。

下面以用户和订单之间的关系为例讲解一对多级联查询（实现"根据用户id查询用户及其关联的订单信息"的功能）的处理过程，读者只需参考该实例即可学会一对多级联查询的MyBatis实现。

【例13-8】 一对多级联查询的MyBatis实现。

该实例是在【例13-7】的基础上实现的，相同的部分不再赘述。其他的具体实现如下。

1. 创建数据表

本实例需要两张数据表：一张是用户表user，一张是订单表orders。这两张表具有一对多的级联关系。前面已创建了user表，orders的创建代码如下：

```
CREATE TABLE 'orders' (
  'id' int(11) NOT NULL AUTO_INCREMENT,
  'ordersn' varchar(10) COLLATE utf8_unicode_ci DEFAULT NULL,
  'user_id' int(11) DEFAULT NULL,
```

```
    PRIMARY KEY ('id'),
    KEY 'user_id' ('user_id'),
    CONSTRAINT 'user_id' FOREIGN KEY ('user_id') REFERENCES 'user' ('uid')
);
```

2. 创建持久化类

在 ch13_2 应用的 com.po 包中创建数据表 orders 对应的持久化类 Orders，及创建数据表 user 对应的持久化类 MyUser。

MyUser 类的代码如下：

```
package com.po;
import java.util.List;
public class MyUser {
    private Integer uid;//主键
    private String uname;
    private String usex;
    private List<Orders>ordersList;
    //省略 setter 和 getter 方法
    @Override
    public String toString() {
    return "User [uid=" +uid +",uname=" +uname +",usex=" +usex +",ordersList=" + ordersList +"]";
    }
}
```

Orders 类的代码如下：

```
package com.po;
/**
 *springtest 数据库中 orders 表的持久化类
 */
public class Orders {
    private Integer id;
    private  String ordersn;
    //省略 setter 和 getter 方法
    @Override
    public String toString() {
        return "Orders [id=" +id +",ordersn=" +ordersn +"]";
    }
}
```

3. 创建映射文件

在 ch13_2 的 com.mybatis.mapper 中创建两张表对应的映射文件 UserMapper.xml

和OrdersMapper.xml。在映射文件UserMapper.xml中实现一对多级联查询（根据用户id查询用户及其关联的订单信息）。

UserMapper.xml的代码如下：

```xml
<?xml version="1.0" encoding="UTF-8"?>
<!DOCTYPE mapper
PUBLIC "-//mybatis.org//DTD Mapper 3.0//EN"
"http://mybatis.org/dtd/mybatis-3-mapper.dtd">
<mapper namespace="com.dao.UserDao">
    <!--一对多 根据uid查询用户及其关联的订单信息:级联查询的第一种方法(嵌套查询) -->
    <resultMap type="MyUser" id="userAndOrders1">
        <id property="uid" column="uid"/>
        <result property="uname" column="uname"/>
        <result property="usex" column="usex"/>
        <!--一对多关联查询,ofType表示集合中的元素类型,将uid传递给selectOrdersById-->
        <collection property="ordersList" ofType="Orders" column="uid"
select="com.dao.OrdersDao.selectOrdersById"/>
    </resultMap>
    <select id="selectUserOrdersById1" parameterType="Integer" resultMap=
    "userAndOrders1">
        select * from user where uid=#{id}
    </select>
    <!--一对多 根据uid查询用户及其关联的订单信息:级联查询的第二种方法(嵌套结果) -->
    <resultMap type="MyUser" id="userAndOrders2">
        <id property="uid" column="uid"/>
        <result property="uname" column="uname"/>
        <result property="usex" column="usex"/>
        <!--一对多关联查询,ofType表示集合中的元素类型 -->
        <collection property="ordersList" ofType="Orders" >
            <id property="id" column="id"/>
            <result property="ordersn" column="ordersn"/>
        </collection>
    </resultMap>
    <select id="selectUserOrdersById2" parameterType="Integer" resultMap=
"userAndOrders2">
        select u.*,o.id,o.ordersn from user u, orders o where u.uid=o.user_id and u.uid=#{id}
    </select>
    <!--一对多 根据uid查询用户及其关联的订单信息:连接查询(使用POJO存储结果) -->
    <select id="selectUserOrdersById3" parameterType="Integer" resultType=
    "SelectUserOrdersById">
        select u.*,o.id,o.ordersn from user u, orders o where u.uid=o.user_id and u.uid=#{id}
    </select>
</mapper>
```

OrdersMapper.xml 的代码如下：

```xml
<?xml version="1.0" encoding="UTF-8" ?>
<!DOCTYPE mapper
PUBLIC "-//mybatis.org//DTD Mapper 3.0//EN"
"http://mybatis.org/dtd/mybatis-3-mapper.dtd">
<mapper namespace="com.dao.OrdersDao">
    <!--根据用户 uid 查询订单信息 -->
    <select id="selectOrdersById" parameterType="Integer" resultType="Orders">
        select * from orders where user_id=#{id}
    </select>
</mapper>
```

4. 创建 POJO 类

在 ch13_2 的 com.pojo 包中创建第 3 步中使用的 POJO 类 SelectUserOrdersById。SelectUserOrdersById 的代码如下：

```java
package com.pojo;
public class SelectUserOrdersById {
    private Integer uid;
    private String uname;
    private String usex;
    private Integer id;
    private String ordersn;
    //省略 setter 和 getter 方法
    @Override
    public String toString() {
        return "User [uid=" +uid +",uname=" +uname +",usex=" +usex +",oid=" +id +",ordersn=" +ordersn +"]";
    }
}
```

5. 在 MyBatis 核心配置文件中指定类型别名和映射文件位置

在 MyBatis 核心配置文件中为 MyUser、Orders 及 SelectUserOrdersById 指定类型别名，代码如下：

```xml
<typeAlias type="com.po.MyUser" alias="MyUser"/>
<typeAlias type="com.po.Orders" alias="Orders"/>
<typeAlias type="com.pojo.SelectUserOrdersById" alias="SelectUserOrdersById"/>
```

在 MyBatis 核心配置文件中指定映射文件位置，代码如下：

```xml
<mapper resource="com/mybatis/mapper/UserMapper.xml"/>
<mapper resource="com/mybatis/mapper/OrdersMapper.xml"/>
```

6. 创建数据操作接口

在 ch13_2 的 com.dao 包中创建第 3 步映射文件对应的数据操作接口 OrdersDao 和 UserDao。

OrdersDao 的代码如下：

```
package com.dao;
import java.util.List;
import org.springframework.stereotype.Repository;
import com.po.Orders;
@Repository("ordersDao")
public interface OrdersDao {
    public List<Orders>selectOrdersById(Integer uid);
}
```

UserDao 的代码如下：

```
package com.dao;
import java.util.List;
import com.po.MyUser;
import com.pojo.SelectUserOrdersById;
@Repository("userDao")
public interface UserDao {
    public MyUser selectUserOrdersById1(Integer uid);
    public MyUser selectUserOrdersById2(Integer uid);
    public List<SelectUserOrdersById>selectUserOrdersById3(Integer uid);
}
```

7. 创建控制器类

在 ch13_2 的 com.controller 包中创建控制器类 OneToMoreController，在该类中调用第 6 步的接口方法。具体代码如下：

```
package com.controller;
import java.util.List;
import org.springframework.beans.factory.annotation.Autowired;
import org.springframework.stereotype.Controller;
import org.springframework.web.bind.annotation.RequestMapping;
import com.dao.UserDao;
import com.po.MyUser;
import com.pojo.SelectUserOrdersById;
@Controller
```

```
public class OneToMoreController {
    @Autowired
    private UserDao userDao;
    @RequestMapping("/oneToMoreTest")
    public String oneToMoreTest() {
        //查询一个用户及订单信息
        System.out.println("级联查询的第一种方法(嵌套查询,执行两个 SQL 语句)");
        MyUser auser1 =userDao.selectUserOrdersById1(1);
        System.out.println(auser1);
        System.out.println("=====================================");
        System.out.println("级联查询的第二种方法(嵌套结果,执行一个 SQL 语句)");
        MyUser auser2 =userDao.selectUserOrdersById2(1);
        System.out.println(auser2);
        System.out.println("=====================================");
        System.out.println("连接查询(使用 POJO 存储结果)");
        List<SelectUserOrdersById>auser3 =userDao.selectUserOrdersById3(1);
        System.out.println(auser3);
        return "test";
    }
}
```

8. 测试应用

重启 Web 服务器 Tomcat,通过地址 http://localhost:8080/ch13_2/oneToMoreTest 测试应用。测试时,需事先为数据表手动添加数据。运行结果如图 13.5 所示。

```
Problems  Declaration  Javadoc  Console  Progress  Servers
Tomcat v9.0 Server at localhost [Apache Tomcat] C:\Program Files\Java\jdk-11.0.6\bin\javaw.exe (2020年2月16日 上午10:15:36)
级联查询的第一种方法(嵌套查询,执行两个SQL语句)
DEBUG [http-nio-8080-exec-8] - ==>  Preparing: select * from user where uid = ?
DEBUG [http-nio-8080-exec-8] - ==> Parameters: 1(Integer)
DEBUG [http-nio-8080-exec-8] - <==      Total: 1
DEBUG [http-nio-8080-exec-8] - ==>  Preparing: select * from orders where user_id=?
DEBUG [http-nio-8080-exec-8] - ==> Parameters: 1(Integer)
DEBUG [http-nio-8080-exec-8] - <==      Total: 3
User [uid=1,uname=张三,usex=女,ordersList=[Orders [id=1,ordersn=12345], Orders [id=2,
=====================================
级联查询的第二种方法(嵌套结果,执行一个SQL语句)
DEBUG [http-nio-8080-exec-8] - ==>  Preparing: select u.*,o.id,o.ordersn from user u,
DEBUG [http-nio-8080-exec-8] - ==> Parameters: 1(Integer)
DEBUG [http-nio-8080-exec-8] - <==      Total: 3
User [uid=1,uname=张三,usex=女,ordersList=[Orders [id=1,ordersn=12345], Orders [id=2,
=====================================
连接查询(使用POJO存储结果)
DEBUG [http-nio-8080-exec-8] - ==>  Preparing: select u.*,o.id,o.ordersn from user u,
DEBUG [http-nio-8080-exec-8] - ==> Parameters: 1(Integer)
DEBUG [http-nio-8080-exec-8] - <==      Total: 3
[User [uid=1,uname=张三,usex=女,oid=1,ordersn=12345], User [uid=1,uname=张三,usex=女,
```

图 13.5　一对多级联查询结果

13.7.3 多对多级联查询

其实,MyBatis没有实现多对多级联,这是因为多对多级联可以通过两个一对多级联进行替换。例如,一个订单可以有多种商品,一种商品可以对应多个订单,订单与商品就是多对多的级联关系。使用一个中间表订单记录表,就可以将多对多级联转换成两个一对多的关系。下面以订单和商品(实现"查询所有订单以及每个订单对应的商品信息"的功能)为例讲解多对多级联查询。

【例13-9】 多对多级联查询的MyBatis实现。

该实例是在【例13-8】的基础上实现的,相同的部分不再赘述。其他的具体实现如下。

1. 创建数据表

前文已创建订单表,这里需要再创建商品表product和订单记录表orders_detail。创建代码如下:

```
CREATE TABLE 'product' (
  'id' int(11) NOT NULL AUTO_INCREMENT,
  'name' varchar(50) COLLATE utf8_unicode_ci DEFAULT NULL,
  'price' double DEFAULT NULL,
  PRIMARY KEY ('id')
);
CREATE TABLE 'orders_detail' (
  'id' int(11) NOT NULL AUTO_INCREMENT,
  'orders_id' int(11) DEFAULT NULL,
  'product_id' int(11) DEFAULT NULL,
  PRIMARY KEY ('id'),
  KEY 'orders_id' ('orders_id'),
  KEY 'product_id' ('product_id'),
  CONSTRAINT 'orders_id' FOREIGN KEY ('orders_id') REFERENCES 'orders' ('id'),
  CONSTRAINT 'product_id' FOREIGN KEY ('product_id') REFERENCES 'product' ('id')
);
```

2. 创建持久化类

在ch13_2应用的com.po包中创建数据表product对应的持久化类Product,而中间表orders_detail不需要持久化类,但需要在订单表orders对应的持久化类Orders中添加关联属性。

Product的代码如下:

```
package com.po;
import java.util.List;
```

```java
public class Product {
    private Integer id;
    private String name;
    private Double price;
    //多对多中的一个一对多
    private List<Orders>orders;
    //省略 setter 和 getter 方法
    @Override
    public String toString() {
        return "Product [id=" +id +",name=" +name +",price=" +price +"]";
    }
}
```

Orders 的代码如下:

```java
package com.po;
import java.util.List;

/**
 *springtest 数据库中 orders 表的持久化类
 */
public class Orders {
    private Integer id;
    private String ordersn;
    //多对多中的另一个一对多
    private List<Product>products;
    //省略 setter 和 getter 方法
    @Override
    public String toString() {
        return "Orders [id=" +id +",ordersn=" +ordersn +",products=" +products +"]";
    }
}
```

3. 创建映射文件

本实例只需在 com.mybatis.mapper 的 OrdersMapper.xml 文件中追加以下配置,即可实现多对多关联查询。

```xml
<!--多对多关联 查询所有订单以及每个订单对应的商品信息(嵌套结果) -->
<resultMap type="Orders" id="allOrdersAndProducts">
    <id property="id" column="id"/>
    <result property="ordersn" column="ordersn"/>
    <!--多对多关联 -->
    <collection property="products" ofType="com.po.Product">
        <id property="id" column="pid"/>
```

```
            <result property="name" column="name"/>
            <result property="price" column="price"/>
        </collection>
</resultMap>
<select id="selectallOrdersAndProducts" resultMap="allOrdersAndProducts">
    select o.*,p.id as pid,p.name,p.price
    from orders o,orders_detail od,product p
    where od.orders_id =o.id
    and od.product_id =p.id
</select>
```

4. 创建数据操作接口方法

在 Orders 接口中添加以下接口方法:

```
public List<Orders> selectallOrdersAndProducts();
```

5. 创建控制器类

在 ch13_2 的 com.controller 包中创建类 MoreToMoreController,在该类中调用第 4 步的接口方法。

MoreToMoreController 的代码如下:

```
package com.controller;
import java.util.List;
import org.springframework.beans.factory.annotation.Autowired;
import org.springframework.web.bind.annotation.RequestMapping;
import com.dao.OrdersDao;
import com.po.Orders;
@Controller
public class MoreToMoreController {
    @Autowired
    private OrdersDao ordersDao;
    @RequestMapping("/moreToMoreTest")
    public String test() {
        List<Orders> os =ordersDao.selectallOrdersAndProducts();
        for (Orders orders : os) {
            System.out.println(orders);
        }
        return "test";
    }
}
```

6. 测试应用

重启 Web 服务器 Tomcat,通过地址 http://localhost:8080/ch13_2/moreToMoreTest 测

试应用。测试时,需事先为数据表手动添加数据。运行结果如图 13.6 所示。

```
DEBUG [http-nio-8080-exec-5] - ==>  Preparing: select o.*,p.id as pid,p.name,p.price from orders 
DEBUG [http-nio-8080-exec-5] - ==> Parameters: 
DEBUG [http-nio-8080-exec-5] - <==      Total: 3
Orders [id=1,ordersn=12345,products=[Product [id=1,name=苹果,price=10.8], Product [id=2,name=香蕉,
Orders [id=2,ordersn=54321,products=[Product [id=1,name=苹果,price=10.8]]]
```

图 13.6　多对多级联查询结果

13.8　动态 SQL

开发人员通常根据需求手动拼接 SQL 语句,这是一个极其麻烦的工作,而 MyBatis 提供了对 SQL 语句动态组装的功能,恰能解决这一问题。MyBatis 的动态 SQL 元素和使用 JSTL 或其他类似基于 XML 的文本处理器相似,常用元素有＜if＞＜choose＞＜when＞＜otherwise＞＜trim＞＜where＞＜set＞＜foreach＞和＜bind＞等元素。

13.8.1　＜if＞元素

动态 SQL 通常要做的事情是有条件地包含 where 子句的一部分。所以在 MyBatis 中,＜if＞元素是最常用的元素,它类似于 Java 中的 if 语句。下面通过一个实例讲解＜if＞元素的使用过程。

【例 13-10】　＜if＞元素的使用过程。

该实例是在【例 13-1】的基础上实现的,相同的实现不再赘述。其他的具体实现如下。

1. 添加 Mapper 接口方法

在应用 ch13_1 的 UserMapper 接口中添加数据操作接口方法 selectAllUserByIf(),在该方法中使用 MyUser 类的对象封装参数信息。接口方法 selectAllUserByIf()的定义如下:

```
public List<MyUser>selectAllUserByIf(MyUser user);
```

2. 添加 SQL 映射

在 SQL 映射文件 UserMapper.xml 中添加接口方法 selectAllUserByIf()对应的 SQL 映射,具体代码如下:

```xml
<!--使用 if 元素,根据条件动态查询用户信息 -->
<select id="selectAllUserByIf"  resultType="MyUser" parameterType="MyUser">
    select * from user where 1=1
    <if test="uname !=null and uname!=''">
        and uname like concat('%',#{uname},'%')
```

```
        </if>
        <if test="usex !=null and usex!=''">
            and usex = #{usex}
        </if>
</select>
```

3. 添加请求处理方法

在控制器类 UserController 中添加请求处理方法 selectAllUserByIf()，具体代码如下：

```
/**
 * 使用 if 元素,根据条件动态查询用户信息
 */
@RequestMapping("/selectAllUserByIf")
public String selectAllUserByIf(Model model) {
    MyUser mu = new MyUser();
    mu.setUname("陈");
    mu.setUsex("男");
    List<MyUser> unameAndUsexList = userMapper.selectAllUserByIf(mu);
    model.addAttribute("unameAndUsexList", unameAndUsexList);
    return "showUnameAndUsexUser";
}
```

4. 测试应用

重启 Web 服务器 Tomcat，通过地址 http://localhost:8080/ch13_1/selectAllUserByIf 测试应用。

13.8.2 ＜choose＞＜when＞＜otherwise＞元素

有时不需要用到所有的条件语句，而只需从中择其一二。针对这种情况，MyBatis 提供了＜choose＞元素，它有点像 Java 中的 switch 语句。下面通过一个实例讲解＜choose＞元素的使用过程。

【例 13-11】 ＜choose＞元素的使用过程。

该实例是在【例 13-1】的基础上实现的，相同的实现不再赘述。其他的具体实现如下。

1. 添加 Mapper 接口方法

在应用 ch13_1 的 UserMapper 接口中添加数据操作接口方法 selectUserByChoose()，在该方法中使用 MyUser 类的对象封装参数信息。接口方法 selectUserByChoose() 的定义如下：

```
public List<MyUser> selectUserByChoose(MyUser user);
```

2. 添加 SQL 映射

在 SQL 映射文件 UserMapper.xml 中添加接口方法 selectUserByChoose() 对应的 SQL 映射，具体代码如下：

```xml
<!--使用 choose、when、otherwise 元素，根据条件动态查询用户信息 -->
<select id="selectUserByChoose" resultType="MyUser" parameterType="MyUser">
    select * from user where 1=1
    <choose>
        <when test="uname !=null and uname!=''">
            and uname like concat('%',#{uname},'%')
        </when>
        <when test="usex !=null and usex!=''">
            and usex =#{usex}
        </when>
        <otherwise>
            and uid >3
        </otherwise>
    </choose>
</select>
```

3. 添加请求处理方法

在控制器类 UserController 中添加请求处理方法 selectUserByChoose()，具体代码如下：

```java
/**
 * 使用 choose、when、otherwise 元素，根据条件动态查询用户信息
 */
@RequestMapping("/selectUserByChoose")
public String selectUserByChoose(Model model) {
    MyUser mu =new MyUser();
    mu.setUname("");
    mu.setUsex("");
    List<MyUser>unameAndUsexList =userMapper.selectUserByChoose(mu);
    model.addAttribute("unameAndUsexList", unameAndUsexList);
    return "showUnameAndUsexUser";
}
```

4. 测试应用

重启 Web 服务器 Tomcat，通过地址 http://localhost:8080/ch13_1/selectUserByChoose 测试应用。

13.8.3 ＜trim＞元素

＜trim＞元素的主要功能是可以在自己包含的内容前加上某些前缀，也可以在其后加上某些后缀，与之对应的属性是 prefix 和 suffix。可以把包含内容首部的某些内容覆盖掉，即忽略，也可以把尾部的某些内容覆盖掉，对应的属性是 prefixOverrides 和 suffixOverrides。正因为＜trim＞元素有这样的功能，所以也可以非常简单地利用＜trim＞来代替＜where＞元素的功能。下面通过一个实例讲解＜trim＞元素的使用过程。

【例 13-12】 ＜trim＞元素的使用过程。

该实例是在【例 13-1】的基础上实现的，相同的实现不再赘述。其他的具体实现如下：

1. 添加 Mapper 接口方法

在应用 ch13_1 的 UserMapper 接口中添加数据操作接口方法 selectUserByTrim()，在该方法中使用 MyUser 类的对象封装参数信息。接口方法 selectUserByTrim() 的定义如下：

```
public List<MyUser> selectUserByTrim(MyUser user);
```

2. 添加 SQL 映射

在 SQL 映射文件 UserMapper.xml 中添加接口方法 selectUserByTrim() 对应的 SQL 映射，具体代码如下：

```xml
<!--使用 trim 元素,根据条件动态查询用户信息 -->
<select id="selectUserByTrim"  resultType="MyUser" parameterType="MyUser">
    select * from user
    <trim prefix="where" prefixOverrides="and|or">
        <if test="uname !=null and uname!=''">
            and uname like concat('%',#{uname},'%')
        </if>
        <if test="usex !=null and usex!=''">
            and usex = #{usex}
        </if>
    </trim>
</select>
```

3. 添加请求处理方法

在控制器类 UserController 中添加请求处理方法 selectUserByTrim()，具体代码如下：

```
/**
```

```
 * 使用trim元素查询用户信息
 */
@RequestMapping("/selectUserByTrim")
public String selectUserByTrim(Model model) {
    MyUser mu = new MyUser();
    mu.setUname("陈");
    mu.setUsex("男");
    List<MyUser>unameAndUsexList = userMapper.selectUserByTrim(mu);
    model.addAttribute("unameAndUsexList", unameAndUsexList);
    return "showUnameAndUsexUser";
}
```

4．测试应用

重启 Web 服务器 Tomcat，通过地址 http://localhost:8080/ch13_1/selectUserByTrim 测试应用。

13.8.4 ＜where＞元素

＜where＞元素的作用是输出一个 where 语句，优点是不考虑 ＜where＞元素的条件输出，MyBatis 将智能处理。如果所有的条件都不满足，那么 MyBatis 将会查出所有记录，如果输出是 and 开头，MyBatis 将忽略第一个 and，如果是 or 开头，MyBatis 也将忽略它；此外，在 ＜where＞元素中不考虑空格的问题，MyBatis 将智能加上。下面通过一个实例讲解 ＜where＞元素的使用过程。

【例 13-13】 ＜where＞元素的使用过程。

该实例是在【例 13-1】的基础上实现的，相同的实现不再赘述。其他的具体实现如下。

1．添加 Mapper 接口方法

在应用 ch13_1 的 UserMapper 接口中添加数据操作接口方法 selectUserByWhere()，在该方法中使用 MyUser 类的对象封装参数信息。接口方法 selectUserByWhere() 的定义如下：

```
public List<MyUser>selectUserByWhere(MyUser user);
```

2．添加 SQL 映射

在 SQL 映射文件 UserMapper.xml 中添加接口方法 selectUserByWhere() 对应的 SQL 映射，具体代码如下：

```
<!--使用 where 元素，根据条件动态查询用户信息 -->
<select id="selectUserByWhere" resultType="MyUser" parameterType="MyUser">
```

```
            select * from user
            <where>
                <if test="uname !=null and uname!=''">
                    and uname like concat('%',#{uname},'%')
                </if>
                <if test="usex !=null and usex!=''">
                    and usex =#{usex}
                </if>
            </where>
        </select>
```

3. 添加请求处理方法

在控制器类 UserController 中添加请求处理方法 selectUserByWhere()，具体代码如下：

```
/**
 * 使用 where 元素查询用户信息
 */
@RequestMapping("/selectUserByWhere")
public String selectUserByWhere(Model model) {
    MyUser mu =new MyUser();
    mu.setUname("陈");
    mu.setUsex("男");
    List<MyUser>unameAndUsexList =userMapper.selectUserByWhere(mu);
    model.addAttribute("unameAndUsexList", unameAndUsexList);
    return "showUnameAndUsexUser";
}
```

4. 测试应用

重启 Web 服务器 Tomcat，通过地址 http://localhost:8080/ch13_1/selectUserByWhere 测试应用。

13.8.5 <set>元素

在 update 语句中可以使用<set>元素动态更新列。下面通过一个实例讲解<set>元素的使用过程。

【例 13-14】 <set>元素的使用过程。

该实例是在【例 13-1】的基础上实现的，相同的实现不再赘述。其他的具体实现如下。

1. 添加 Mapper 接口方法

在应用 ch13_1 的 UserMapper 接口中添加数据操作接口方法 updateUserBySet()，

在该方法中使用 MyUser 类的对象封装参数信息。接口方法 updateUserBySet() 的定义如下：

```
public int updateUserBySet(MyUser user);
```

2. 添加 SQL 映射

在 SQL 映射文件 UserMapper.xml 中添加接口方法 updateUserBySet() 对应的 SQL 映射，具体代码如下：

```xml
<!--使用 set 元素,动态修改一个用户 -->
<update id="updateUserBySet" parameterType="MyUser">
    update user
    <set>
        <if test="uname !=null">uname=#{uname},</if>
        <if test="usex !=null">usex=#{usex}</if>
    </set>
    where uid =#{uid}
</update>
```

3. 添加请求处理方法

在控制器类 UserController 中添加请求处理方法 updateUserBySet()，具体代码如下：

```java
/**
 * 使用 set 元素,动态修改一个用户
 */
@RequestMapping("/updateUserBySet")
public String updateUserBySet(Model model) {
    MyUser setmu =new MyUser();
    setmu.setUid(3);
    setmu.setUname("张九");
    userMapper.updateUserBySet(setmu);
    //查询出来看看 id 为 3 的用户是否被修改
    List<Map<String, Object>>unameAndUsexList =userMapper.selectAllUserMap();
    model.addAttribute("unameAndUsexList", unameAndUsexList);
    return "showUnameAndUsexUser";
}
```

4. 测试应用

重启 Web 服务器 Tomcat，通过地址 http://localhost:8080/ch13_1/updateUserBySet 测试应用。

13.8.6 ＜foreach＞元素

＜foreach＞元素主要用于构建 in 条件，它可以在 SQL 语句中进行迭代一个集合。＜foreach＞元素的属性主要有 item、index、collection、open、separator、close。item 表示集合中每一个元素进行迭代时的别名，index 指定一个名字，用于表示在迭代过程中每次迭代到的位置，open 表示该语句以什么开始，separator 表示每次进行迭代以什么符号作为分隔符，close 表示以什么结束。使用＜foreach＞时，最关键也是最容易出错的是 collection 属性，该属性是必选的，但在不同情况下，该属性的值是不一样的，主要有以下 3 种情况：

- 如果传入的是单参数且参数类型是一个 list 的时候，collection 的属性值为 list。
- 如果传入的是单参数且参数类型是一个 array 数组的时候，collection 的属性值为 array。
- 如果传入的参数是多个，需要把它们封装成一个 Map，当然单参数也可以封装成 Map。Map 的 key 是参数名，所以 collection 属性值是传入的 List 或 array 对象在自己封装的 Map 中的 key。

下面通过一个实例讲解＜set＞元素的使用过程。

【例 13-15】 ＜foreach＞元素的使用过程。

该实例是在【例 13-1】的基础上实现的，相同的实现不再赘述。其他的具体实现如下。

1. 添加 Mapper 接口方法

在应用 ch13_1 的 UserMapper 接口中添加数据操作接口方法 selectUserByForeach()，在该方法中使用 List 作为参数。接口方法 selectUserByForeach()的定义如下：

```
public List<MyUser>selectUserByForeach(List<Integer>listId);
```

2. 添加 SQL 映射

在 SQL 映射文件 UserMapper.xml 中添加接口方法 selectUserByForeach()对应的 SQL 映射，具体代码如下：

```xml
<!--使用 foreach 元素,查询用户信息 -->
<select id="selectUserByForeach" resultType="MyUser" parameterType="List">
    select * from user where uid in
    <foreach item="item" index="index" collection="list"
    open="(" separator="," close=")">
        #{item}
    </foreach>
</select>
```

3. 添加请求处理方法

在控制器类 UserController 中添加请求处理方法 selectUserByForeach()，具体代码如下：

```java
/**
 * 使用 foreach 元素,查询用户信息
 */
@RequestMapping("/selectUserByForeach")
public String selectUserByForeach(Model model) {
    List<Integer> listId = new ArrayList<Integer>();
    listId.add(4);
    listId.add(5);
    List<MyUser> unameAndUsexList = userMapper.selectUserByForeach(listId);
    model.addAttribute("unameAndUsexList", unameAndUsexList);
    return "showUnameAndUsexUser";
}
```

4. 测试应用

重启 Web 服务器 Tomcat,通过地址 http://localhost:8080/ch13_1/selectUserByForeach 测试应用。

13.8.7 <bind>元素

在模糊查询时,如果使用"${}"拼接字符串,无法防止 SQL 注入问题。如果使用字符串拼接函数或连接符号,但不同数据库的拼接函数或连接符号不同,如 MySQL 的 concat 函数、Oracle 的连接符号"||"。这样,SQL 映射文件就需要根据不同的数据库提供不同的实现,显然比较麻烦,且不利于代码的移植。幸运的是,MyBatis 提供了<bind>元素来解决这一问题。

下面通过一个实例讲解<bind>元素的使用过程。

【例 13-16】 <bind>元素的使用过程。

该实例是在【例 13-1】的基础上实现的,相同的实现不再赘述。其他的具体实现如下。

1. 添加 Mapper 接口方法

在应用 ch13_1 的 UserMapper 接口中添加数据操作接口方法 selectUserByBind()。接口方法 selectUserByBind()的定义如下:

```java
public List<MyUser> selectUserByBind(MyUser user);
```

2. 添加 SQL 映射

在 SQL 映射文件 UserMapper.xml 中添加接口方法 selectUserByBind()对应的 SQL 映射,具体代码如下:

```xml
<!--使用bind元素进行模糊查询 -->
<select id="selectUserByBind" resultType="MyUser" parameterType="MyUser">
    <!--bind中 uname 是 com.po.MyUser 的属性名 -->
```

```
        <bind name="paran_uname" value="'%' +uname +'%'"/>
        select * from user where uname like #{paran_uname}
</select>
```

3. 添加请求处理方法

在控制器类 UserController 中添加请求处理方法 selectUserByBind(),具体代码如下:

```
/**
 * 使用 bind 元素查询用户信息
 */
@RequestMapping("/selectUserByBind")
public String selectUserByBind(Model model) {
    MyUser bindmu =new MyUser();
    bindmu.setUname("陈");
    List<MyUser> unameAndUsexList =userMapper.selectUserByBind(bindmu);
    model.addAttribute("unameAndUsexList", unameAndUsexList);
    return "showUnameAndUsexUser";
}
```

4. 测试应用

重启 Web 服务器 Tomcat,通过地址 http://localhost:8080/ch13_1/selectUserByBind 测试应用。

13.9 本章小结

本章首先简要介绍 MyBatis 的核心配置文件,这是因为核心配置文件不会轻易改动;其次详细讲解了 SQL 映射文件的主要元素,包括＜select＞＜resultMap＞等主要元素;最后以实例为主讲解了级联查询的 MyBatis 实现,读者参考这些实例即可实现自己需要的级联查询。

习 题 13

1. MyBatis 实现查询时,返回的结果集有几种常见的存储方式?请举例说明。
2. 在 MyBatis 中,针对不同的数据库软件,＜insert＞元素如何将主键回填?
3. 在 MyBatis 中,如何给 SQL 语句传递参数?
4. 在动态 SQL 元素中,类似分支语句的元素有哪些?如何使用它们?

名片管理系统的设计与实现

学习目的与要求

本章通过名片管理系统的设计与实现,讲述如何使用 Spring MVC+MyBatis 框架来实现一个 Web 应用。通过本章的学习,掌握 Spring MVC+MyBatis 框架应用开发的流程、方法以及技术。

本章主要内容

- 系统设计。
- 数据库设计。
- 系统管理。
- 组件设计。
- 系统实现。

本章系统使用 Spring MVC+MyBatis 框架实现各个模块,Web 引擎为 Tomcat 9.0,数据库采用的是 MySQL 5.x,集成开发环境为 STS 或 Eclipse。

14.1 系统设计

14.1.1 系统功能需求

名片管理系统是针对注册用户使用的系统。系统提供的功能如下：
（1）非注册用户可以注册为注册用户。
（2）成功注册的用户，可以登录系统。
（3）成功登录的用户，可以添加、修改、删除以及浏览自己客户的名片信息。
（4）成功登录的用户，可以修改密码。

14.1.2 系统模块划分

用户登录成功后，进入管理主页面（selectCards.jsp），可以对自己的客户名片进行管理。系统模块划分如图14.1所示。

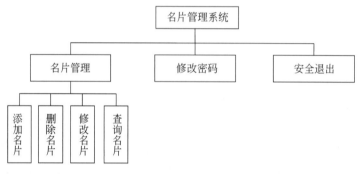

图14.1　名片管理系统

14.2 数据库设计

系统采用加载纯Java数据库驱动程序的方式连接MySQL5.x数据库。在MySQL5.x的数据库card14中，共创建两张与系统相关的数据表：usertable和cardtable。

14.2.1 数据库概念结构设计

根据系统设计与分析，可以设计出如下数据结构：

1. 用户

包括ID、用户名以及密码，注册用户名唯一。

2. 名片

包括ID、名称、电话、邮箱、单位、职务、地址、Logo以及所属用户。其中，ID唯一，"所属用户"与"1. 用户"的用户 ID 关联。

根据以上数据结构,结合数据库设计特点,可画出图 14.2 所示的数据库概念结构图。

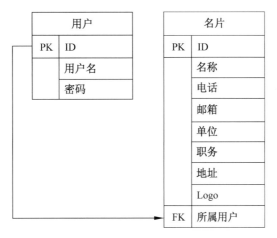

图 14.2　数据库概念结构

其中,ID 为正整数,值是从 1 开始递增的序列。

14.2.2　数据库逻辑结构设计

将数据库概念结构图转换为 MySQL 数据库所支持的实际数据模型,即数据库的逻辑结构。

用户信息表(usertable)的设计如表 14.1 所示。

表 14.1　用户信息

字　段	含　义	类　型	长　度	是否为空
id	编号(PK)	int	11	no
uname	用户名	varchar	50	no
upwd	密码	varchar	32	no

名片信息表(cardtable)的设计如表 14.2 所示。

表 14.2　名片信息

字　段	含　义	类　型	长　度	是否为空
id	编号(PK)	int	11	no
name	名称	varchar	50	no
telephone	电话	varchar	20	no
email	邮箱	varchar	50	
company	单位	varchar	50	
post	职务	varchar	50	

续表

字段	含义	类型	长度	是否为空
address	地址	varchar	50	
logoName	图片	varchar	30	
user_id	所属用户	int	11	no

14.3 系统管理

14.3.1 Maven 项目依赖管理

使用 STS 创建一个名为 ch14 的 Maven Project，并通过 pom.xml 文件添加项目所依赖的 JAR 包（包括 MyBatis、Spring MVC、Spring JDBC、MySQL 连接器、MyBatis 与 Spring 桥接器、Log4j、Fileupload、Jackson、DBCP 以及 JSTL 等依赖）。ch14 的 pom.xml 文件内容如下：

```xml
<project xmlns="http://maven.apache.org/POM/4.0.0"
    xmlns:xsi="http://www.w3.org/2001/XMLSchema-instance"
    xsi:schemaLocation="http://maven.apache.org/POM/4.0.0
https://maven.apache.org/xsd/maven-4.0.0.xsd">
    <modelVersion>4.0.0</modelVersion>
    <groupId>com.cn.chenheng</groupId>
    <artifactId>ch14</artifactId>
    <version>0.0.1-SNAPSHOT</version>
    <packaging>war</packaging>
    <properties>
        <!--spring 版本号 -->
        <spring.version>5.2.3.RELEASE</spring.version>
    </properties>
    <dependencies>
        <dependency>
            <groupId>org.springframework</groupId>
            <artifactId>spring-webmvc</artifactId>
            <version>${spring.version}</version>
        </dependency>
        <dependency>
            <groupId>org.springframework</groupId>
            <artifactId>spring-jdbc</artifactId>
            <version>${spring.version}</version>
        </dependency>
        <dependency>
            <groupId>mysql</groupId>
```

```xml
            <artifactId>mysql-connector-java</artifactId>
            <version>5.1.45</version>
        </dependency>
        <dependency>
            <groupId>org.apache.commons</groupId>
            <artifactId>commons-dbcp2</artifactId>
            <version>2.7.0</version>
        </dependency>
        <dependency>
            <groupId>org.mybatis</groupId>
            <artifactId>mybatis</artifactId>
            <version>3.5.4</version>
        </dependency>
        <dependency>
            <groupId>org.mybatis</groupId>
            <artifactId>mybatis-spring</artifactId>
            <version>2.0.3</version>
        </dependency>
        <dependency>
            <groupId>log4j</groupId>
            <artifactId>log4j</artifactId>
            <version>1.2.17</version>
        </dependency>
        <dependency>
            <groupId>javax.servlet</groupId>
            <artifactId>jstl</artifactId>
            <version>1.2</version>
        </dependency>
        <dependency>
            <groupId>commons-fileupload</groupId>
            <artifactId>commons-fileupload</artifactId>
            <version>1.4</version>
        </dependency>
        <dependency>
            <groupId>com.fasterxml.jackson.core</groupId>
            <artifactId>jackson-databind</artifactId>
            <version>2.10.2</version>
        </dependency>
    </dependencies>
</project>
```

14.3.2 JSP 页面管理

为方便管理，在 src/main/webapp/static 目录下存放与系统相关的静态资源，如

BootStrap 相关的 CSS 与 JS；在 src/main/webapp/WEB-INF/jsp 目录下存放与系统相关的 JSP 页面。由于篇幅所限，本章仅附上 JSP 和 Java 文件的代码，具体代码请参考本书提供的源代码 ch14。

1. 首页面

在 src/main/webapp 目录下创建应用的首页面 index.jsp，首页面重定向到 user/toLogin 请求，打开登录页面。index.jsp 的代码如下：

```jsp
<%@page language="java" contentType="text/html; charset=UTF-8" pageEncoding="UTF-8"%>
<%
String path = request.getContextPath();
String basePath = request.getScheme()+"://"+request.getServerName()+":"+request.getServerPort()+path+"/";
%>
<!DOCTYPE html>
<html>
<head>
<base href="<%=basePath%>">
<meta charset="UTF-8">
<title>Insert title here</title>
</head>
<body>
    <%response.sendRedirect("user/toLogin");%>
</body>
</html>
```

2. 异常信息显示页面

本系统使用 Spring 框架的统一异常处理机制处理未登录异常和程序错误异常。为显示异常信息，需在 src/main/webapp/WEB-INF/jsp 目录下创建一个名为 error.jsp 的页面。error.jsp 的代码如下：

```jsp
<%@page language="java" contentType="text/html; charset=UTF-8" isErrorPage="true"
pageEncoding="UTF-8"%>
<%@taglib prefix="c" uri="http://java.sun.com/jsp/jstl/core" %>
<%
String path = request.getContextPath();
String basePath = request.getScheme() +"://" + request.getServerName() +":" + request.getServerPort() +path + "/";
%>
<!DOCTYPE html>
<html>
```

```
<head>
<base href="<%=basePath%>">
<meta charset="UTF-8">
<title>Insert title here</title>
<link href="static/css/bootstrap.min.css" rel="stylesheet">
</head>
<body>
    <%
        exception.printStackTrace(response.getWriter());
    %>
    <c:if test="${mymessage == 'noLogin'}">
        <h2>没登录,您没有权限访问,请<a href="user/toLogin">登录</a>!</h2>
    </c:if>
    <c:if test="${mymessage == 'noError'}">
        <h2>服务器内部错误或资源不存在!</h2>
    </c:if>
</body>
</html>
```

14.3.3 包管理

本系统的包层次结构如图 14.3 所示。

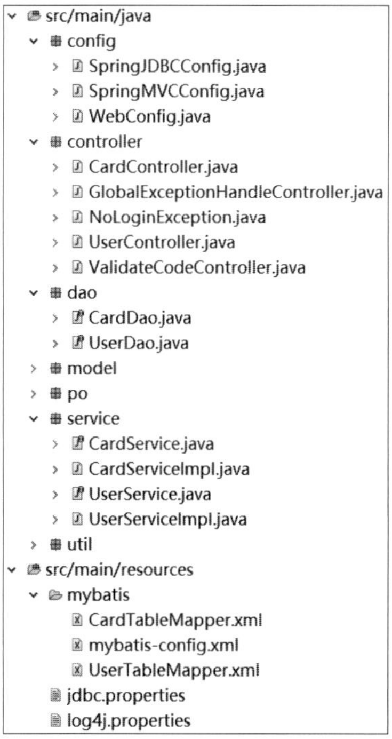

```
▼ ⊞ src/main/java
    ▼ ⊞ config
        ▷ ⓙ SpringJDBCConfig.java
        ▷ ⓙ SpringMVCConfig.java
        ▷ ⓙ WebConfig.java
    ▼ ⊞ controller
        ▷ ⓙ CardController.java
        ▷ ⓙ GlobalExceptionHandleController.java
        ▷ ⓙ NoLoginException.java
        ▷ ⓙ UserController.java
        ▷ ⓙ ValidateCodeController.java
    ▼ ⊞ dao
        ▷ ⓙ CardDao.java
        ▷ ⓙ UserDao.java
    ▷ ⊞ model
    ▷ ⊞ po
    ▼ ⊞ service
        ▷ ⓙ CardService.java
        ▷ ⓙ CardServiceImpl.java
        ▷ ⓙ UserService.java
        ▷ ⓙ UserServiceImpl.java
    ▷ ⊞ util
▼ ⊞ src/main/resources
    ▼ ⊟ mybatis
            ⓧ CardTableMapper.xml
            ⓧ mybatis-config.xml
            ⓧ UserTableMapper.xml
        ⓟ jdbc.properties
        ⓟ log4j.properties
```

图 14.3 包层次结构

1. config 包

该包存放的类是系统的配置类，包括 Web 配置类、Spring MVC 配置类以及 Spring JDBC 配置类。

2. controller 包

该包存放的类是系统的控制器类和异常处理类，包括名片管理相关的控制器类、用户相关的控制器类、验证码控制器类以及全局异常处理类。

3. dao 包

该包存放的 Java 程序是 @Mapper 注解的数据操作接口。包括名片管理相关的数据访问接口和用户相关的数据访问接口。

4. model 包

该包存放的类是两个领域模型类，与表单对应：Card 封装名片信息和 MyUser 封装用户信息。

5. po 包

该包存放的类是两个持久化类，与两个数据表对应。

6. service 包

该包存放的类是业务处理类，是控制器和 dao 的桥梁。包下有 Service 接口和 Service 实现类。

7. util 包

该包存放的类是工具类，包括 MyUtil 类（文件重命名）和 MD5Util 类（MD5 加密）。

8. mybatis 包

该包中存放的文件是 MyBatis 的全局配置文件和 SQL 映射文件。

14.3.4 配置类管理

名片管理系统共有 3 个配置类，分别是 Web 的配置类 WebConfig、Spring MVC 的配置类 SpringMVCConfig 和 Spring JDBC 的配置类 SpringJDBCConfig。

在 SpringMVCConfig 类中配置视图解析器、静态资源以及上传文件的相关设置，具体代码如下：

```
package config;
import java.io.IOException;
import org.springframework.context.annotation.Bean;
```

```java
import org.springframework.context.annotation.ComponentScan;
import org.springframework.context.annotation.Configuration;
import org.springframework.core.io.FileSystemResource;
import org.springframework.core.io.Resource;
import org.springframework.web.multipart.commons.CommonsMultipartResolver;
import org.springframework.web.servlet.config.annotation.EnableWebMvc;
import org.springframework.web.servlet.config.annotation.ResourceHandlerRegistry;
import org.springframework.web.servlet.config.annotation.WebMvcConfigurer;
import org.springframework.web.servlet.view.InternalResourceViewResolver;
@Configuration
@EnableWebMvc // 开启 Spring MVC 的支持
@ComponentScan(basePackages = { "controller", "service"}) // 扫描基本包
public class SpringMVCConfig implements WebMvcConfigurer {
    /**
     * 配置视图解析器
     */
    @Bean
    public InternalResourceViewResolver getViewResolver() {
        InternalResourceViewResolver viewResolver = new
            InternalResourceViewResolver();
        viewResolver.setPrefix("/WEB-INF/jsp/");
        viewResolver.setSuffix(".jsp");
        return viewResolver;
    }
    /**
     * 配置静态资源(不需要 DispatcherServlet 转发的请求)
     */
    @Override
    public void addResourceHandlers(ResourceHandlerRegistry registry) {
        registry.addResourceHandler("/static/**").addResourceLocations
("/static/");
    }
    /**
     * 配置上传文件的相关设置
     */
    @Bean("multipartResolver")
    public CommonsMultipartResolver getMultipartResolver() {
        CommonsMultipartResolver cmr = new CommonsMultipartResolver();
        //设置请求的编码格式,默认为 iso-8859-1
        cmr.setDefaultEncoding("UTF-8");
        //设置允许上传文件的最大值,单位为字节
        cmr.setMaxUploadSize(5400000);
        //设置上传文件的临时路径
```

```
            //workspace\.metadata\.plugins\org.eclipse.wst.server.core\tmp0\
wtpwebapps\fileUpload
        Resource uploadTempDir = new FileSystemResource("fileUpload/temp");
        try {
            cmr.setUploadTempDir(uploadTempDir);
        } catch (IOException e) {
            // TODO Auto-generated catch block
            e.printStackTrace();
        }
        return cmr;
    }
}
```

在SpringJDBCConfig类中配置数据源、为数据源添加事务管理器以及配置MyBatis工厂,具体代码如下:

```
package config;
import org.apache.commons.dbcp2.BasicDataSource;
import org.mybatis.spring.SqlSessionFactoryBean;
import org.mybatis.spring.annotation.MapperScan;
import org.springframework.beans.factory.annotation.Value;
import org.springframework.context.annotation.Bean;
import org.springframework.context.annotation.Configuration;
import org.springframework.context.annotation.PropertySource;
import org.springframework.core.io.ClassPathResource;
import org.springframework.core.io.Resource;
import org.springframework.jdbc.datasource.DataSourceTransactionManager;
import org.springframework.transaction.annotation.EnableTransactionManagement;
@Configuration // 通过该注解来表明该类是一个Spring的配置,相当于一个xml文件
@PropertySource(value = { "classpath:jdbc.properties" }, ignoreResourceNotFound = true)
//配置多个配置文件  value={"classpath:jdbc.properties","xx","xxx"}
@EnableTransactionManagement // 开启声明式事务的支持
@MapperScan(basePackages = { "dao" }) //配置扫描MyBatis接口的包路径
public class SpringJDBCConfig {
    @Value("${db.url}") // 注入属性文件jdbc.properties中的jdbc.url
    private String jdbcUrl;
    @Value("${db.driverClassName}")
    private String jdbcDriverClassName;
    @Value("${db.username}")
    private String jdbcUsername;
    @Value("${db.password}")
    private String jdbcPassword;
    @Value("${db.maxTotal}")
```

```java
    private int maxTotal;
    @Value("${db.maxIdle}")
    private int maxIdle;
    @Value("${db.initialSize}")
    private int initialSize;
    /**
     * 配置数据源
     */
    @Bean
    public BasicDataSource dataSource() {
        BasicDataSource myDataSource = new BasicDataSource();
        // 数据库驱动
        myDataSource.setDriverClassName(jdbcDriverClassName);
        // 相应驱动的jdbcUrl
        myDataSource.setUrl(jdbcUrl);
        // 数据库的用户名
        myDataSource.setUsername(jdbcUsername);
        // 数据库的密码
        myDataSource.setPassword(jdbcPassword);
        // 最大连接数
        myDataSource.setMaxTotal(maxTotal);
        // 最大空闲连接数
        myDataSource.setMaxIdle(maxIdle);
        // 初始化连接数
        myDataSource.setInitialSize(initialSize);
        return myDataSource;
    }
    /**
     * 为数据源添加事务管理器
     */
    @Bean
    public DataSourceTransactionManager transactionManager() {
        DataSourceTransactionManager dt = new DataSourceTransactionManager();
        dt.setDataSource(dataSource());
        return dt;
    }
    /**
     * 配置MyBatis工厂,同时指定数据源,并与MyBatis完美整合
     */
    @Bean
    public SqlSessionFactoryBean getSqlSession() {
        SqlSessionFactoryBean sqlSessionFactory = new SqlSessionFactoryBean();
        sqlSessionFactory.setDataSource(dataSource());
        Resource r = new ClassPathResource("mybatis/mybatis-config.xml");
```

```java
            sqlSessionFactory.setConfigLocation(r);
            return sqlSessionFactory;
    }
}
```

在WebConfig类中注册Spring MVC的配置类SpringMVCConfig、Spring JDBC的配置类SpringJDBCConfig、Spring MVC的DispatcherServlet以及字符编码过滤器,具体代码如下:

```java
package config;
import javax.servlet.ServletContext;
import javax.servlet.ServletException;
import javax.servlet.ServletRegistration.Dynamic;
import org.springframework.web.WebApplicationInitializer;
import org.springframework.web.context.support.AnnotationConfigWebApplicationContext;
import org.springframework.web.filter.CharacterEncodingFilter;
import org.springframework.web.servlet.DispatcherServlet;
public class WebConfig implements WebApplicationInitializer{
    @Override
    public void onStartup(ServletContext arg0) throws ServletException {
        AnnotationConfigWebApplicationContext ctx =new 
        AnnotationConfigWebApplicationContext();
        ctx.register(SpringJDBCConfig.class);//注册配置类SpringJDBCConfig
        ctx.register(SpringMVCConfig.class);//注册Spring MVC的Java配置类
                                             //SpringMVCConfig
        ctx.setServletContext(arg0);//和当前ServletContext关联
        /**
         * 注册Spring MVC的DispatcherServlet
         */
        Dynamic servlet = arg0.addServlet("dispatcher", new DispatcherServlet(ctx));
        servlet.addMapping("/");
        servlet.setLoadOnStartup(1);
        /**
         * 注册字符编码过滤器
         */
        javax.servlet.FilterRegistration.Dynamic filter =arg0.addFilter
        ("characterEncodingFilter",
        CharacterEncodingFilter.class);
        filter.setInitParameter("encoding", "UTF-8");
        filter.addMappingForUrlPatterns(null, false, "/*");
    }
}
```

14.3.5 配置文件管理

名片管理系统的配置文件包括日志配置文件 log4j.properties、数据库连接信息配置文件 jdbc.properties 以及 MyBatis 全局配置文件 mybatis-config.xml。

日志配置文件 log4j.properties 的内容如下：

```
#Global logging configuration
log4j.rootLogger=ERROR, stdout
#MyBatis logging configuration...
log4j.logger.dao=DEBUG
#Console output...
log4j.appender.stdout=org.apache.log4j.ConsoleAppender
log4j.appender.stdout.layout=org.apache.log4j.PatternLayout
log4j.appender.stdout.layout.ConversionPattern=%5p [%t] -%m%n
```

数据库连接信息配置文件 jdbc.properties 的内容如下：

```
db.driverClassName=com.mysql.jdbc.Driver
db.url=jdbc:mysql://localhost:3306/card14?characterEncoding=utf8
db.username=root
db.password=root
db.maxTotal=30
db.maxIdle=15
db.initialSize=5
```

在 MyBatis 全局配置文件 mybatis-config.xml 中配置实体类别名以及 SQL 映射文件位置，具体内容如下：

```
<?xml version="1.0" encoding="UTF-8" ?>
<!DOCTYPE configuration
PUBLIC "-//mybatis.org//DTD Config 3.0//EN"
"http://mybatis.org/dtd/mybatis-3-config.dtd">
<configuration>
    <settings>
        <!--输出日志配置 -->
        <setting name="logImpl" value="LOG4J" />
    </settings>
    <typeAliases>
        <typeAlias type="model.MyUser" alias="MyUser" />
        <typeAlias type="po.MyUserTable" alias="MyUserTable" />
        <typeAlias type="model.Card" alias="Card" />
        <typeAlias type="po.CardTable" alias="CardTable" />
    </typeAliases>
    <!--告诉 MyBatis 到哪里去找映射文件 -->
    <mappers>
```

```xml
        <mapper resource="mybatis/UserTableMapper.xml"/>
        <mapper resource="mybatis/CardTableMapper.xml"/>
    </mappers>
</configuration>
```

14.4 组件设计

名片管理系统的组件包括工具类、统一异常处理类和验证码类。

14.4.1 工具类

名片管理系统的工具类包括 MyUtil 和 MD5Util。在 MyUtil 类中定义一个文件重命名方法 getNewFileName，MyUtil 类的代码如下：

```java
package util;
import java.text.SimpleDateFormat;
import java.util.Date;
public class MyUtil {
    /**
     * 将实际的文件名重命名
     */
    public static String getNewFileName(String oldFileName) {
        int lastIndex = oldFileName.lastIndexOf(".");
        String fileType = oldFileName.substring(lastIndex);
        Date now = new Date();
        SimpleDateFormat sdf = new SimpleDateFormat("YYYYMMDDHHmmssSSS");
        String time = sdf.format(now);
        String newFileName = time + fileType;
        return newFileName;
    }
}
```

在 MD5Util 类中定义了 MD5 加密方法，具体代码如下：

```java
package util;
import java.security.MessageDigest;
public class MD5Util {
    /***
     * MD5加码生成32位 md5码
     */
    public static String string2MD5(String inStr) {
        MessageDigest md5 = null;
        try {
            md5 = MessageDigest.getInstance("MD5");
        } catch (Exception e) {
```

```java
        e.printStackTrace();
        return "";
    }
    char[] charArray = inStr.toCharArray();
    byte[] byteArray = new byte[charArray.length];
    for (int i = 0; i < charArray.length; i++)
        byteArray[i] = (byte) charArray[i];
    byte[] md5Bytes = md5.digest(byteArray);
    StringBuffer hexValue = new StringBuffer();
    for (int i = 0; i < md5Bytes.length; i++) {
        int val = ((int) md5Bytes[i]) & 0xff;
        if (val < 16)
            hexValue.append("0");
        hexValue.append(Integer.toHexString(val));
    }
    return hexValue.toString();
}
/***
 * 自己规则加密
 * @param inStr
 * @return
 */
public static String MD5(String inStr){
    String xy = "abc";
    String finalStr="";
    if(inStr!=null){
        String fStr = inStr.substring(0, 1);
        String lStr = inStr.substring(1, inStr.length());
        finalStr = string2MD5( fStr+xy+lStr);
    }else{
        finalStr = string2MD5(xy);
    }
    return finalStr;
}
}
```

14.4.2 统一异常处理

名片管理系统采用@ControllerAdvice注解实现异常的统一处理，统一处理了NoLoginException和Exception异常，具体代码如下：

```java
package controller;
import org.springframework.ui.Model;
import org.springframework.web.bind.annotation.ControllerAdvice;
```

```java
import org.springframework.web.bind.annotation.ExceptionHandler;
/**
 * 统一异常处理
 */
@ControllerAdvice
public class GlobalExceptionHandleController {
    @ExceptionHandler(value=Exception.class)
    public String exceptionHandler(Exception e, Model model) {
        String message ="";
        if (e instanceof NoLoginException) {
            message ="noLogin";
        } else {//未知异常
            message =  "noError";
        }
        model.addAttribute("mymessage",message);
        return "error";
    }
}
```

未登录异常类 NoLoginException 的代码如下：

```java
package controller;
public class NoLoginException extends Exception{
    private static final long serialVersionUID =1L;
    public NoLoginException() {
        super();
    }
    public NoLoginException(String message) {
        super(message);
    }
}
```

14.4.3 验证码

本系统验证码的使用步骤如下：

1. 创建产生验证码的控制器类

在 controller 包中创建产生验证码的控制器类 ValidateCodeController，具体代码如下：

```java
package controller;
import java.awt.Color;
import java.awt.Font;
import java.awt.Graphics;
import java.awt.image.BufferedImage;
```

```java
import java.io.IOException;
import java.io.OutputStream;
import java.util.Random;
import javax.imageio.ImageIO;
import javax.servlet.ServletException;
import javax.servlet.http.HttpServletRequest;
import javax.servlet.http.HttpServletResponse;
import javax.servlet.http.HttpSession;
import org.springframework.stereotype.Controller;
import org.springframework.web.bind.annotation.RequestMapping;
@Controller
public class ValidateCodeController {
    private char code[] = {'a', 'b', 'c', 'd', 'e', 'f', 'g', 'h', 'i', 'j','k', 'm',
                    'n', 'p', 'q', 'r', 's', 't', 'u', 'v', 'w', 'x', 'y','z',
                    'A', 'B', 'C', 'D', 'E', 'F', 'G', 'H', 'J', 'K', 'L', 'M',
                    'N', 'P', 'Q', 'R', 'S', 'T', 'U', 'V', 'W', 'X', 'Y', 'Z',
                    '2','3', '4', '5', '6', '7', '8', '9' };
    private static final int WIDTH = 50;
    private static final int HEIGHT = 20;
    private static final int LENGTH = 4;
    @RequestMapping("/validateCode")
    public void validateCode(HttpServletRequest request,
HttpServletResponse response) throws ServletException, IOException {
        // TODO Auto-generated method stub
        // 设置响应报头信息
        response.setHeader("Pragma", "No-cache");
        response.setHeader("Cache-Control", "no-cache");
        response.setDateHeader("Expires", 0);
        // 设置响应的 MIME 类型
        response.setContentType("image/jpeg");
        BufferedImage image = new BufferedImage(WIDTH, HEIGHT,
BufferedImage.TYPE_INT_RGB);
        Font mFont = new Font("Arial", Font.TRUETYPE_FONT, 18);
        Graphics g = image.getGraphics();
        Random rd = new Random();
        // 设置背景颜色
        g.setColor(new Color(rd.nextInt(55) +200, rd.nextInt(55) +200,
rd.nextInt(55) +200));
        g.fillRect(0, 0, WIDTH, HEIGHT);
        // 设置字体
        g.setFont(mFont);
        // 画边框
        g.setColor(Color.black);
        g.drawRect(0, 0, WIDTH -1, HEIGHT -1);
```

```java
// 随机产生的验证码
String result = "";
for (int i = 0; i < LENGTH; ++i) {
    result += code[rd.nextInt(code.length)];
}
HttpSession se = request.getSession();
se.setAttribute("rand", result);
// 画验证码
for (int i = 0; i < result.length(); i++) {
    g.setColor(new Color(rd.nextInt(200), rd.nextInt(200), rd
            .nextInt(200)));
    g.drawString(result.charAt(i) + "", 12 * i + 1, 16);
}
// 随机产生2个干扰线
for (int i = 0; i < 2; i++) {
    g.setColor(new Color(rd.nextInt(200), rd.nextInt(200), rd
            .nextInt(200)));
    int x1 = rd.nextInt(WIDTH);
    int x2 = rd.nextInt(WIDTH);
    int y1 = rd.nextInt(HEIGHT);
    int y2 = rd.nextInt(HEIGHT);
    g.drawLine(x1, y1, x2, y2);
}
// 释放图形资源
g.dispose();
try {
    OutputStream os = response.getOutputStream();
    // 输出图像到页面
    ImageIO.write(image, "JPEG", os);
} catch (IOException e) {
    e.printStackTrace();
}
    }
}
```

2. 使用验证码

在需要使用验证码的JSP页面中调用产生验证码的控制器显示验证码,示例代码片段如下:

```html
<div class="form-group has-success">
    <label class="col-sm-2 col-md-2 control-label">验证码</label>
    <div class="col-sm-4 col-md-4">
        <table style="width: 100%">
```

```html
            <tr>
                <td>
<form:input cssClass="form-control" placeholder="请输入验证码" path="code"/>
</td>
                <td>
                    <img src="validateCode" id="mycode">
                </td>
                <td>
                    <a href="javascript:refreshCode()">换一张</a>
                </td>
            </tr>
        </table>
    </div>
</div>
```

14.5 名片管理

14.5.1 领域模型与持久化类

在本系统中,领域模型简单地作为视图对象,它的作用是将某个指定页面的所有数据封装起来,与表单对应。数据传递方向为 View→Controller→Service→Dao。

与名片管理相关的领域模型是 Card(位于 model 包),具体代码如下:

```java
package model;
import org.springframework.web.multipart.MultipartFile;
public class Card {
    private int id;
    private String name;
    private String telephone;
    private String email;
    private String company;
    private String post;
    private String address;
    private MultipartFile logo;
    private String logoName;
    private int user_id;
    //省略 set 和 get 方法
}
```

在本系统中,持久层是关系型数据库,所以,持久化类的每个属性对应数据表中的每个字段。数据传递方向为 Dao →Service→Controller→View。

与名片管理相关的持久化类是 CardTable(位于 po 包),具体代码如下:

```java
package po;
```

```
public class CardTable {
    private int id;
    private String name;
    private String telephone;
    private String email;
    private String company;
    private String post;
    private String address;
    private String logoName;
    private int user_id;
    //省略 set 和 get 方法
}
```

14.5.2　Controller 实现

在本系统中，与名片管理相关的功能包括添加、修改、删除、查询等，由控制器类 CardController 负责处理。由系统功能需求可知，用户必须成功登录才能管理自己的名片，所以，CardController 处理添加、修改、删除、查询名片等功能前，需要进行登录权限验证。在 CardController 中，使用 @ModelAttribute 注解的方法进行登录权限验证。CardController 的具体代码如下：

```
package controller;
import java.io.IOException;
import javax.servlet.http.HttpServletRequest;
import javax.servlet.http.HttpSession;
import org.springframework.beans.factory.annotation.Autowired;
import org.springframework.stereotype.Controller;
import org.springframework.ui.Model;
import org.springframework.web.bind.annotation.ModelAttribute;
import org.springframework.web.bind.annotation.RequestMapping;
import org.springframework.web.bind.annotation.ResponseBody;
import model.Card;
import model.MyUser;
import service.CardService;
@Controller
@RequestMapping("/card")
public class CardController {
    @Autowired
    private CardService cardService;
    /**
     * 权限控制
     */
    @ModelAttribute
    public void checkLogin(HttpSession session) throws NoLoginException{
```

```java
        if(session.getAttribute("userLogin") ==null) {
            throw new NoLoginException();
        }
}
/**
 * 查询、修改查询、删除查询
 */
@RequestMapping("/selectAllCardsByPage")
public String selectAllCardsByPage(Model model, int currentPage, String
act, HttpSession session) {
        return cardService.selectAllCardsByPage(model, currentPage, act,
        session);
}
/**
 * 打开添加页面
 */
@RequestMapping("/toAddCard")
public String toAddCard(@ModelAttribute Card card) {
    return "addCard";
}
/**
 * 实现添加及修改功能
 */
@RequestMapping("/addCard")
 public String addCard (@ ModelAttribute Card card, HttpServletRequest
 request, String act, HttpSession session) throws IllegalStateException,
 IOException {
    return cardService.addCard(card, request, act, session);
}
/**
 * 打开详情及修改页面
 */
@RequestMapping("/detail")
public String detail(Model model, int id, String act) {
    return cardService.detail(model, id, act);
}
/**
 * 删除
 */
@RequestMapping("/delete")
@ResponseBody
public String delete(int id) {
    return cardService.delete(id);
}
```

```java
/**
 * 安全退出
 */
@RequestMapping("/loginOut")
public String loginOut(Model model, HttpSession session) {
    return cardService.loginOut(model, session);
}
/**
 * 打开修改密码页面
 */
@RequestMapping("/toUpdatePwd")
public String toUpdatePwd(Model model, HttpSession session) {
    return cardService.toUpdatePwd(model, session);
}
/**
 * 修改密码
 */
@RequestMapping("/updatePwd")
public String updatePwd(@ModelAttribute MyUser myuser) {
    return cardService.updatePwd(myuser);
}
}
```

14.5.3 Service 实现

与名片管理相关的 Service 接口和实现类分别为 CardService 和 CardServiceImpl。控制器获取一个请求后，需要调用 Service 层中业务处理方法，在 Service 层中需要调用 Dao 层。所以，Service 层是控制器层和 Dao 层的桥梁。

CardService 接口的代码如下：

```java
package service;
import java.io.IOException;
import javax.servlet.http.HttpServletRequest;
import javax.servlet.http.HttpSession;
import org.springframework.ui.Model;
import model.Card;
import model.MyUser;
public interface CardService {
    public String selectAllCardsByPage(Model model, int currentPage, String
        act, HttpSession session);
    public String addCard(Card card, HttpServletRequest request, String act,
        HttpSession session) throws IllegalStateException, IOException;
    public String detail(Model model, int id, String act);
    public String delete(int id);
```

```java
    public String loginOut(Model model, HttpSession session);
    public String toUpdatePwd(Model model, HttpSession session);
    public String updatePwd(MyUser myuser);
}
```

CardServiceImpl 实现类的代码如下：

```java
package service;
import java.io.File;
import java.io.IOException;
import java.util.List;
import java.util.Map;
import javax.servlet.http.HttpServletRequest;
import javax.servlet.http.HttpSession;
import org.springframework.beans.factory.annotation.Autowired;
import org.springframework.stereotype.Service;
import org.springframework.ui.Model;
import org.springframework.web.multipart.MultipartFile;
import util.MD5Util;
import util.MyUtil;
import dao.CardDao;
import model.Card;
import model.MyUser;
import po.CardTable;
import po.MyUserTable;
@Service
public class CardServiceImpl implements CardService{
    @Autowired
    private CardDao cardDao;
    /**
     * 查询、修改查询、删除查询、分页查询
     */
    @Override
    public String selectAllCardsByPage(Model model, int currentPage, String
    act, HttpSession session) {
        MyUserTable mut = (MyUserTable)session.getAttribute("userLogin");
        List<Map<String, Object>> allUser = cardDao.selectAllCards(mut.getId());
        //共多少个用户
        int totalCount = allUser.size();
        //计算共多少页
        int pageSize = 5;
        int totalPage = (int)Math.ceil(totalCount * 1.0/pageSize);
        List<Map<String, Object>> cardsByPage =
                cardDao.selectAllCardsByPage((currentPage-1) * pageSize, pageSize,
                mut.getId());
```

```java
        model.addAttribute("act", act);
        model.addAttribute("allCards", cardsByPage);
        model.addAttribute("totalPage", totalPage);
        model.addAttribute("currentPage", currentPage);
        return "selectCards";
    }
    /**
     * 添加与修改名片
     */
    @Override
    public String addCard(Card card, HttpServletRequest request, String act, HttpSession session) throws IllegalStateException, IOException {
        MultipartFile myfile = card.getLogo();
        //如果选择了上传文件,将文件上传到指定的目录 static/images
        if(!myfile.isEmpty()) {
            //上传文件路径(生产环境)
            String path = request.getServletContext().getRealPath("/static/images/");
            //获得上传文件原名
            String fileName = myfile.getOriginalFilename();
            //对文件重命名
            String fileNewName = MyUtil.getNewFileName(fileName);
            File filePath = new File(path + File.separator + fileNewName);
            //如果文件目录不存在,创建目录
            if(!filePath.getParentFile().exists()) {
                filePath.getParentFile().mkdirs();
            }
            //将上传文件保存到一个目标文件中
            myfile.transferTo(filePath);
            //将重命名后的图片名存到 card 对象中,添加时使用
            card.setLogoName(fileNewName);
        }
        if("add".equals(act)) {
            MyUserTable mut = (MyUserTable)session.getAttribute("userLogin");
            card.setUser_id(mut.getId());
            int n = cardDao.addCard(card);
            if(n > 0) //成功
                return " redirect:/card/selectAllCardsByPage? currentPage = 1&act=select";
            //失败
            return "addCard";
        }else {//修改
            int n = cardDao.updateCard(card);
            if(n > 0) //成功
```

```java
            return " redirect:/card/selectAllCardsByPage? currentPage =
                    1&act=updateSelect";
        //失败
        return "updateCard";
    }
}
/**
 * 打开详情与修改页面
 */
@Override
public String detail(Model model, int id, String act) {
    CardTable ct =cardDao.selectACard(id);
    model.addAttribute("card", ct);
    if("detail".equals(act)) {
        return "cardDetail";
    }else {
        return "updateCard";
    }
}
/**
 * 删除
 */
@Override
public String delete(int id) {
    cardDao.deleteACard(id);
    return "/card/selectAllCardsByPage? currentPage=1&act=deleteSelect";
}
/**
 * 安全退出
 */
@Override
public String loginOut(Model model, HttpSession session) {
    session.invalidate();
    model.addAttribute("myUser", new MyUser());
    return "login";
}
/**
 * 打开修改密码页面
 */
@Override
public String toUpdatePwd(Model model, HttpSession session) {
    MyUserTable mut =(MyUserTable)session.getAttribute("userLogin");
    model.addAttribute("myuser", mut);
    return "updatePwd";
```

```java
    }
    /**
     * 修改密码
     */
    @Override
    public String updatePwd(MyUser myuser) {
        //将明文变成密文
        myuser.setUpwd(MD5Util.MD5(myuser.getUpwd()));
        cardDao.updatePwd(myuser);
        return "login";
    }
}
```

14.5.4 Dao 实现

Dao 层是数据访问层,即@Repository 注解的数据操作接口(接口中的方法与 SQL 映射文件中元素的 id 对应),与名片管理相关的数据访问层为 CardDao,具体代码如下:

```java
package dao;
import java.util.List;
import java.util.Map;
import org.apache.ibatis.annotations.Param;
import org.springframework.stereotype.Repository;
import model.Card;
import model.MyUser;
import po.CardTable;
@Repository
public interface CardDao {
    public List<Map<String, Object>> selectAllCards(int uid);
    public List<Map<String, Object>> selectAllCardsByPage(@Param("startIndex")
        int startIndex, @Param("perPageSize") int perPageSize, @Param("uid")
        int uid);
    public int addCard(Card card);
    public int updateCard(Card card);
    public CardTable selectACard(int id);
    public int deleteACard(int id);
    public int updatePwd(MyUser myuser);
}
```

14.5.5 SQL 映射文件

SQL 映射文件的 namespace 属性与数据操作接口对应。与名片管理功能相关的

SQL 映射文件是 CardTableMapper.xml(位于 mybatis 包中),具体代码如下:

```xml
<?xml version="1.0" encoding="UTF-8" ?>
<!DOCTYPE mapper
PUBLIC "-//mybatis.org//DTD Mapper 3.0//EN"
"http://mybatis.org/dtd/mybatis-3-mapper.dtd">
<mapper namespace="dao.CardDao">
    <!--查询所有名片 -->
    <select id="selectAllCards"  resultType="map">
        select * from cardtable where user_id=#{uid}
    </select>
    <!--分页查询名片 -->
    <select id="selectAllCardsByPage" resultType="map">
        select * from cardtable where user_id=#{uid} limit #{startIndex},
    #{perPageSize}
    </select>
    <!--添加名片 -->
    <insert id="addCard" parameterType="Card">
        insert into cardtable (id, name, telephone, email, company, post,
    address,
    logoName, user_id)
         values (null, #{name}, #{telephone}, #{email}, #{company}, #{post},
    #{address}, #{logoName}, #{user_id})
    </insert>
    <!--修改名片 -->
    <update id="updateCard" parameterType="Card">
        update cardtable set
            name=#{name},
            telephone=  #{telephone},
            email=#{email},
            company=#{company},
            post=#{post},
            address=#{address},
            logoName=#{logoName}
        where id=#{id}
    </update>
    <!--查询一个名片,修改及详情使用 -->
    <select id="selectACard" parameterType="integer" resultType="CardTable">
        select * from cardtable where id=#{id}
    </select>
    <!--删除一个名片 -->
    <delete id="deleteACard" parameterType="integer">
        delete from cardtable where id=#{id}
    </delete>
```

```xml
<!--修改密码 -->
<update id="updatePwd" parameterType="myuser">
    update usertable set upwd =#{upwd} where id =#{id}
</update>
</mapper>
```

14.5.6 添加名片

首先，用户登录成功后，进入名片管理系统的主页面。然后，用户在"名片管理"菜单中选择"添加名片"菜单项，打开添加名片页面。最后，用户输入客户名片的姓名、电话、E-mail、单位、职务、地址、Logo，单击"添加"按钮实现添加。如果成功，则跳转到查询视图；如果失败，则回到添加视图。

addCard.jsp 页面实现添加名片信息的输入界面，如图 14.4 所示。

图 14.4 添加名片页面

addCard.jsp 的代码如下：

```
<%@page language="java" contentType="text/html; charset=UTF-8" pageEncoding="UTF-8"%>
<%@taglib prefix="form" uri="http://www.springframework.org/tags/form" %>
<%
String path =request.getContextPath();
String basePath = request.getScheme()+"://"+ request.getServerName()+":"+ request.getServerPort()+path+"/";
%>
<!DOCTYPE html>
```

```html
<html>
<head>
<base href="<%=basePath%>">
<meta charset="UTF-8">
<title>Insert title here</title>
<link rel="stylesheet" href="static/css/bootstrap.min.css" />
</head>
<body>
    <!--加载 header.jsp -->
    <div>
        <jsp:include page="header.jsp"></jsp:include>
    </div>
    <br><br><br>
    <div class="container">
        <div class="bg-primary"  style="width:70%; height: 60px;padding-top: 0.5px;">
            <h3 align="center">添加名片</h3>
      </div><br>
            <form:form action="card/addCard? act=add" method="post" class="form-horizontal" modelAttribute="card" enctype="multipart/form-data" >
                <div class="form-group has-success">
                    <label class="col-sm-2 col-md-2 control-label">姓名</label>
                    <div class="col-sm-4 col-md-4">
                        <form:input cssClass="form-control" placeholder="请输入姓名" path="name"/>
                    </div>
                </div>
                <div class="form-group has-success">
                    <label class="col-sm-2 col-md-2 control-label">电话号码</label>
                    <div class="col-sm-4 col-md-4">
                        <form:input cssClass="form-control" placeholder="请输入电话" path="telephone"/>
                    </div>
                </div>
                <div class="form-group has-success">
                    <label class="col-sm-2 col-md-2 control-label">E-Mail</label>
                    <div class="col-sm-4 col-md-4">
                        <form:input cssClass="form-control" placeholder="请输入 E-Mail" path="email"/>
                    </div>
                </div>
                <div class="form-group has-success">
                    <label class="col-sm-2 col-md-2 control-label">单位</label>
                    <div class="col-sm-4 col-md-4">
```

```html
            <form:input cssClass="form-control" placeholder="请输入单位" path="company"/>
        </div>
    </div>
    <div class="form-group has-success">
        <label class="col-sm-2 col-md-2 control-label">职位</label>
        <div class="col-sm-4 col-md-4">
            <form:input cssClass="form-control" placeholder="请输入职位" path="post"/>
        </div>
    </div>
    <div class="form-group has-success">
        <label class="col-sm-2 col-md-2 control-label">地址</label>
        <div class="col-sm-4 col-md-4">
            <form:input cssClass="form-control" placeholder="请输入地址" path="address"/>
        </div>
    </div>
    <div class="form-group has-success">
        <label class="col-sm-2 col-md-2 control-label">照片</label>
        <div class="col-sm-4 col-md-4">
            <input type="file" placeholder="请选择客户照片" name="logo" class="form-control" />
        </div>
    </div>
    <div class="form-group">
        <div class="col-sm-offset-2 col-sm-10">
            <button type="submit" class="btn btn-success">添加</button>
            <button type="reset" class="btn btn-primary">重置</button>
        </div>
    </div>
</form:form>
</div>
</body>
</html>
```

单击图 14.4 中"添加"按钮,将添加请求通过 card/addCard? act=add 提交给控制器类 CardController(14.5.2 节)的 addCard 方法,进行添加功能处理。添加成功跳转到查询视图;添加失败回到添加视图。

14.5.7 查询名片

用户登录成功后,进入名片管理系统的主页面,主页面中初始显示查询视图 selectCards.jsp,查询页面运行效果如图 14.5 所示。

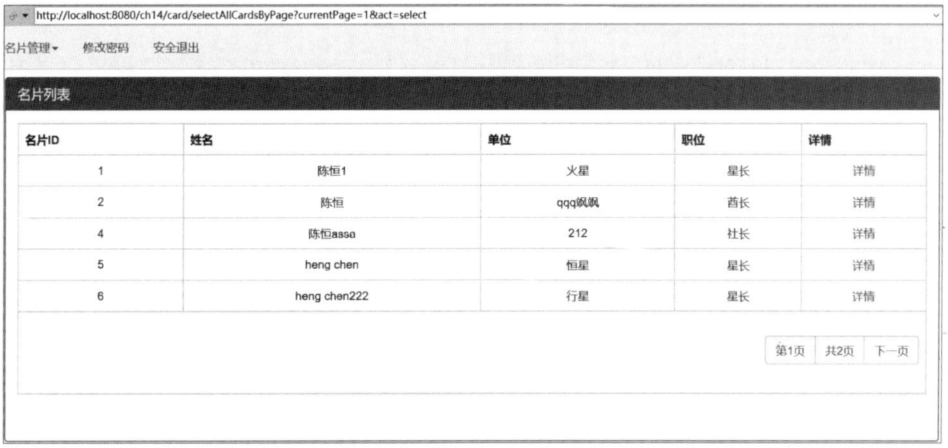

图 14.5 查询页面

单击主页面中"名片管理"菜单的"查询名片"菜单项,打开查询页面 selectCards.jsp。"查询名片"菜单项超链接的目标地址是个 url 请求。该请求路径为 card/selectAllCardsByPage?currentPage=1&act=select,根据请求路径找到对应控制器类 CardController 的 selectAllCardsByPage 方法处理查询功能。在该方法中,根据动作类型("修改查询""查询"以及"删除查询"),将查询结果转发到不同视图。

在 selectCards.jsp 页面中单击"详情"超链接,打开名片详细信息页面 detail.jsp。"详情"超链接的目标地址是个 url 请求。该请求路径为 card/detail?id=${card.id}&act=detail。根据请求路径找到对应控制器类 CardController 的 detail 方法,处理查询一个名片功能。根据动作类型("修改"以及"详情"),将查询结果转发到不同视图。名片详细信息页面 cardDetail.jsp 运行效果如图 14.6 所示。

图 14.6 名片详情

selectCards.jsp 的代码如下：

```jsp
<%@page language="java" contentType="text/html; charset=UTF-8" pageEncoding="UTF-8"%>
<%@taglib prefix="c" uri="http://java.sun.com/jsp/jstl/core" %>
<%
String path = request.getContextPath();
String basePath = request.getScheme()+"://"+request.getServerName()+":"+request.getServerPort()+path+"/";
%>
<!DOCTYPE html>
<html>
<head>
<base href="<%=basePath%>">
<meta charset="UTF-8">
<title>Insert title here</title>
<link rel="stylesheet" href="static/css/bootstrap.min.css" />
<script src="js/jquery.min.js"></script>
<script type="text/javascript">
    function deleteCards(tid){
        if(window.confirm("真的删除该名片吗?")){
            $.ajax(
                {
                    //请求路径
                    url : "card/delete",
                    //请求类型
                    type : "post",
                    //data 表示发送的数据
                    data : {
                        id : tid
                    },
                    //成功响应的结果
                    success : function(obj){//obj 响应数据
                        //获取路径
                        var pathName=window.document.location.pathname;
                        //截取,得到项目名称
                        var projectName=pathName.substring(0,pathName.substring(1).indexOf('/')+1);
                        window.location.href =projectName +obj;
                    },
                    error : function() {
                        alert("处理异常!");
                    }
                }
```

```jsp
            );
        }
    }
</script>
</head>
<body>
    <!--加载 header.jsp -->
    <div>
        <jsp:include page="header.jsp"></jsp:include>
    </div>
    <br><br><br>
    <div class="container">
        <div class="panel panel-primary">
            <div class="panel-heading">
                <h3 class="panel-title">名片列表</h3>
            </div>
            <div class="panel-body">
                <div class="table table-responsive">
                    <table class="table table-bordered table-hover">
                        <tbody class="text-center">
                            <tr>
                                <th>名片 ID</th>
                                <th>姓名</th>
                                <th>单位</th>
                                <th>职位</th>
                                <th>详情</th>
                                <c:if test="${act=='updateSelect'}">
                                    <th>操作</th>
                                </c:if>
                                <c:if test="${act=='deleteSelect'}">
                                    <th>操作</th>
                                </c:if>
                            </tr>
                            <c:if test="${totalPage !=0 }">
                                <c:forEach items="${allCards}" var="card">
                                    <tr>
                                        <td>${card.id}</td>
                                        <td>${card.name}</td>
                                        <td>${card.company}</td>
                                        <td>${card.post}</td>
                                        <td>
                            <a href="card/detail?id=${card.id}&act=detail" target="_blank">详情</a>
                                        </td>
```

```
                        <c:if test="${act=='updateSelect'}">
            <td><a href="card/detail?id=${card.id}&act=update">修改
</a></td>
                        </c:if>
                        <c:if test="${act=='deleteSelect'}">
            <td><a href="javascript:deleteCards('${card.id}')" >
删除</a></td>
                        </c:if>
                    </tr>
                </c:forEach>
                <!--查询 -->
                <c:if test="${act=='select'}">
                    <tr>
                        <td colspan="5" align="right">
                            <ul class="pagination">
                    <li><a>第<span>${currentPage}</span>页</a>
</li>
                                <li><a>共<span>${totalPage}</span>页</a>
</li>
                                <li>
                                    <c:if test="${currentPage !=1}">
<a href="card/selectAllCardsByPage?act=select&currentPage=
${currentPage -1}">上一页</a>
                                    </c:if>
                                    <c:if test=" ${ currentPage   ! =
totalPage}">
<a href="card/selectAllCardsByPage?act=select&currentPage=
${currentPage +1}">下一页</a>
                                    </c:if>
                                </li>
                            </ul>
                        </td>
                    </tr>
                </c:if>
                <!--修改查询 -->
                <c:if test="${act=='updateSelect'}">
                    <tr>
                        <td colspan="6" align="right">
                            <ul class="pagination">
                    <li><a>第<span>${currentPage}</span>页</a>
</li>
                        <li><a>共<span>${totalPage}</span>页
</a></li>
                                <li>
```

```html
                                    <c:if test="${currentPage !=1}">
<a href="card/selectAllCardsByPage?act=updateSelect&currentPage=
${currentPage -1}">上一页</a>
                                    </c:if>
                                    <c:if test="${currentPage !=
                                    totalPage}">
<a href="card/selectAllCardsByPage?act=updateSelect&currentPage=
${currentPage +1}">下一页</a>
                                    </c:if>
                                </li>
                            </ul>
                        </td>
                    </tr>
                </c:if>
                <!--删除查询 -->
                <c:if test="${act=='deleteSelect'}">
                    <tr>
                        <td colspan="6" align="right">
                            <ul class="pagination">
                    <li><a>第<span>${currentPage}</span>页</a>
                                </li>
                     <li><a>共<span>${totalPage}</span>页</a>
                                </li>
                                    <li>
                                    <c:if test="${currentPage !=1}">
<a href="card/selectAllCardsByPage?act=deleteSelect&currentPage=
${currentPage -1}">上一页</a>
                                    </c:if>
                                    <c:if test="${currentPage !=totalPage}">
<a href="card/selectAllCardsByPage?act=deleteSelect&currentPage=
${currentPage +1}">下一页</a>
                                    </c:if>
                                </li>
                            </ul>
                        </td>
                    </tr>
                </c:if>
            </c:if>
        </tbody>
    </table>
  </div>
 </div>
 </div>
</div>
```

```
</body>
</html>
```

cardDetail.jsp 的代码如下：

```jsp
<%@page language="java" contentType="text/html; charset=UTF-8" pageEncoding=
"UTF-8"%>
<%
String path = request.getContextPath();
String basePath = request.getScheme()+"://"+request.getServerName()+":"+
request.getServerPort()+path+"/";
%>
<!DOCTYPE html>
<html>
<head>
<base href="<%=basePath%>">
<meta charset="UTF-8">
<title>Insert title here</title>
<link rel="stylesheet" href="static/css/bootstrap.min.css" />
</head>
<body>
    <!--加载 header.jsp -->
    <div>
        <jsp:include page="header.jsp"></jsp:include>
    </div>
    <br><br><br>
    <div class="container">
        <div class="panel panel-primary">
            <div class="panel-heading">
                <h3 class="panel-title">名片详情</h3>
            </div>
            <div class="panel-body">
                <div class="table table-responsive">
                    <table class="table table-bordered table-hover">
                        <tbody class="text-center">
                            <tr>
                                <th>姓名</th>
                                <td>${card.name}</td>
                            </tr>
                            <tr>
                                <th>电话</th>
                                <td>${card.telephone}</td>
                            </tr>
                            <tr>
                                <th>E-Mail</th>
```

```html
                    <td>${card.email}</td>
                </tr>
                <tr>
                    <th>单位</th>
                    <td>${card.company}</td>
                </tr>
                <tr>
                    <th>职位</th>
                    <td>${card.post}</td>
                </tr>
                <tr>
                    <th>地址</th>
                    <td>${card.address}</td>
                </tr>
                <tr>
                    <th>照片</th>
                    <td>
                        <img src="static/images/${card.logoName}"
                            style="height: 50px; width: 50px; display:
                            block;">
                    </td>
                </tr>
            </tbody>
        </table>
    </div>
   </div>
  </div>
 </div>
</body>
</html>
```

14.5.8 修改名片

单击主页面中"名片管理"菜单的"修改名片"菜单项，打开修改查询页面 selectCards.jsp。"修改名片"菜单项超链接的目标地址是 url 请求 card/selectAllCardsByPage?currentPage＝1&act＝updateSelect。找到对应控制器类 CardController 的方法 selectAllCardsByPage，在该方法中，根据动作类型，将查询结果转发给查询视图 selectCards.jsp。

单击 selectCards.jsp 页面中的"修改"超链接，打开修改名片信息页面 updateCard.jsp。"修改"超链接的目标地址是 url 请求 card/detail? id＝${card.id}&act＝update。找到对应控制器类 CardController 的方法 detail，在该方法中，根据动作类型，将查询结果转发给 updateCard.jsp 页面显示。

输入要修改的信息后单击"修改"按钮，将名片信息提交给控制器类，找到对应控制器类 CardController 的方法 addCard，在 addCard 方法中，根据动作类型执行修改的业务处理。修改成功，则进入修改查询页面。修改失败，则回到 updateCard.jsp 页面。

修改查询页面 selectCards.jsp 的运行效果如图 14.7 所示，updateCard.jsp 页面的运行效果如图 14.8 所示。

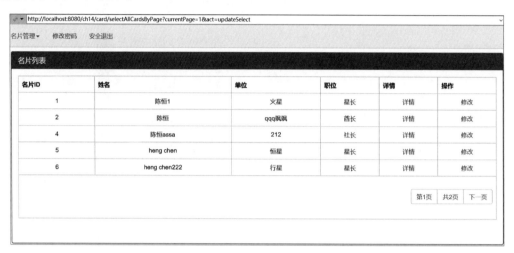

图 14.7 修改查询页面

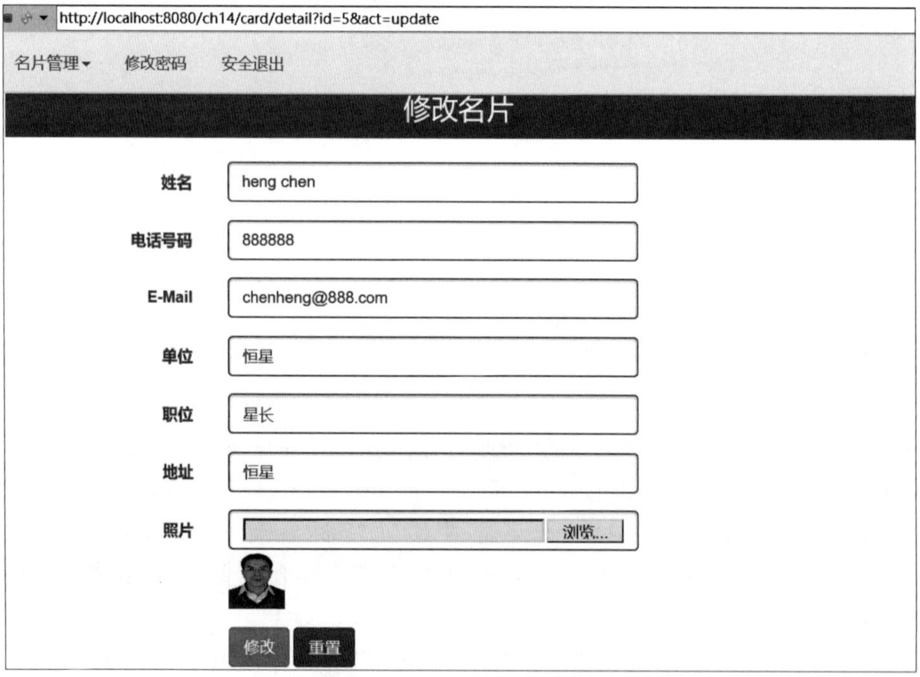

图 14.8 updateCard.jsp 页面

updateCard.jsp 的代码如下：

```
<%@page language="java" contentType="text/html; charset=UTF-8" pageEncoding=
```

```jsp
"UTF-8"%>
<%@taglib prefix="form" uri="http://www.springframework.org/tags/form" %>
<%
String path =request.getContextPath();
String basePath =request.getScheme()+"://"+request.getServerName()+":"+
request.getServerPort()+path+"/";
%>
<!DOCTYPE html>
<html>
<head>
<base href="<%=basePath%>">
<meta charset="UTF-8">
<title>Insert title here</title>
<link rel="stylesheet" href="static/css/bootstrap.min.css" />
</head>
<body>
    <!--加载 header.jsp -->
    <div>
        <jsp:include page="header.jsp"></jsp:include>
    </div>
    <br><br><br>
    <div class="container">
        <div class="bg-primary"   style="width:70%; height: 60px;padding-top:
          0.5px;">
            <h3 align="center">修改名片</h3>
    </div><br>
        <form:form action="card/addCard?act=update" method="post" class=
"form-horizontal" modelAttribute="card" enctype="multipart/form-data" >
            <div class="form-group has-success">
                <label class="col-sm-2 col-md-2 control-label">姓名</label>
                <div class="col-sm-4 col-md-4">
        <form:input   cssClass="form-control" placeholder="请输入姓名" path=
"name"/>
                </div>
            </div>
            <div class="form-group has-success">
                <label class="col-sm-2 col-md-2 control-label">电话号码
                </label>
                <div class="col-sm-4 col-md-4">
        <form:input cssClass="form-control" placeholder="请输入电话" path=
"telephone"/>
                </div>
            </div>
            <div class="form-group has-success">
```

```
                <label class="col-sm-2 col-md-2 control-label">E-Mail</label>
                <div class="col-sm-4 col-md-4">
        <form:input cssClass="form-control" placeholder="请输入 E-Mail" path="email"/>
                </div>
            </div>
            <div class="form-group has-success">
                <label class="col-sm-2 col-md-2 control-label">单位</label>
                <div class="col-sm-4 col-md-4">
        <form:input cssClass="form-control" placeholder="请输入单位" path="company"/>
                </div>
            </div>
            <div class="form-group has-success">
                <label class="col-sm-2 col-md-2 control-label">职位</label>
                <div class="col-sm-4 col-md-4">
        <form:input cssClass="form-control" placeholder="请输入职位" path="post"/>
                </div>
            </div>
            <div class="form-group has-success">
                <label class="col-sm-2 col-md-2 control-label">地址</label>
                <div class="col-sm-4 col-md-4">
        <form:input cssClass="form-control" placeholder="请输入地址" path="address"/>
                </div>
            </div>
            <div class="form-group has-success">
                    <label class="col-sm-2 col-md-2 control-label">照片</label>
                    <div class="col-sm-4 col-md-4">
        <input type="file" placeholder="请选择客户照片" name="logo" class="form-control" />
        <img src="static/images/${card.logoName}" style="height: 50px; width: 50px; display: block;">
                    <form:hidden path="logoName"/>
                    <form:hidden path="id"/>
                </div>
            </div>
            <div class="form-group">
                <div class="col-sm-offset-2 col-sm-10">
                    <button type="submit"class="btn btn-success">修改</button>
                    <button type="reset" class="btn btn-primary">重置</button>
                </div>
            </div>
```

```
            </form:form>
        </div>
    </body>
</html>
```

14.5.9 删除名片

单击主页面中"名片管理"菜单的"删除名片"菜单项,打开删除查询页面 selectCards.jsp。"删除名片"菜单项超链接的目标地址是 url 请求 card/selectAllCardsByPage?currentPage=1&act=deleteSelect。根据请求找到对应控制器类 CardController 的方法 selectAllCardsByPage,在该方法中,根据动作类型,将查询结果转发给 selectCards.jsp 页面,页面效果如图 14.9 所示。

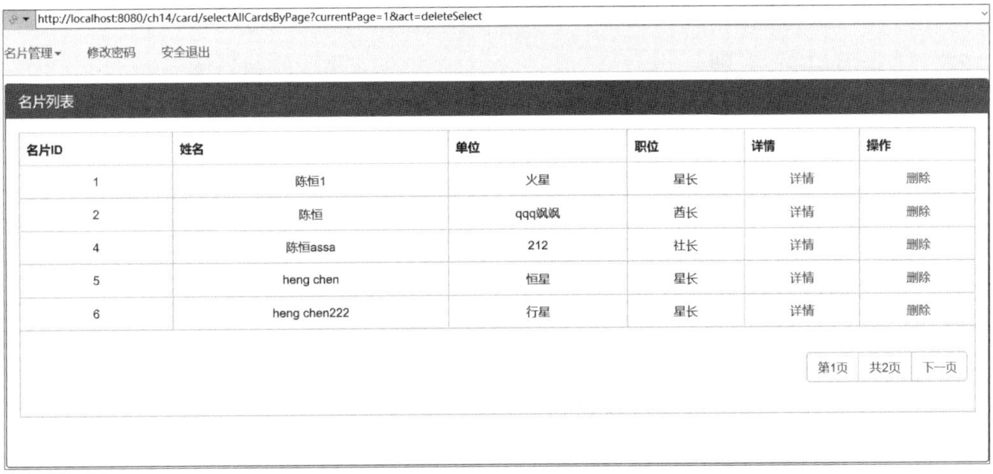

图 14.9 删除查询页面

在图 14.9 中单击"删除"超链接,将要删除名片的 ID 通过 Ajax 提交给控制器类。找到对应控制器类 CardController 的方法 delete,在该方法中执行删除的业务处理。删除成功后,进入删除查询页面。

14.6 用户相关

14.6.1 领域模型与持久化类

与用户相关的领域模型是 MyUser(位于 model 包),具体代码如下:

```
package model;
public class MyUser {
    private int id;
    private String uname;
    private String upwd;
```

```
    private String reupwd;
    private String code;
    //省略 set 和 get 方法
}
```

与用户相关的持久化类是 MyUserTable(位于 po 包),具体代码如下:

```
package po;
public class MyUserTable {
    private Integer id;
    private String uname;
    private String upwd;
    //省略 set 和 get 方法
}
```

14.6.2　Controller 实现

在本系统中,与用户相关的功能包括用户注册、用户登录以及用户检查等,由控制器类 UserController 负责处理。UserController 的具体代码如下:

```
package controller;
import javax.servlet.http.HttpSession;
import org.springframework.beans.factory.annotation.Autowired;
import org.springframework.stereotype.Controller;
import org.springframework.ui.Model;
import org.springframework.web.bind.annotation.ModelAttribute;
import org.springframework.web.bind.annotation.RequestBody;
import org.springframework.web.bind.annotation.RequestMapping;
import org.springframework.web.bind.annotation.ResponseBody;
import model.MyUser;
import service.UserService;
@Controller
@RequestMapping("/user")
public class UserController {
    @Autowired
    private UserService userService;
    @RequestMapping("/toLogin")
    public String toLogin(@ModelAttribute MyUser myUser) {
        return "login";
    }
    @RequestMapping("/toRegister")
    public String toRegister(@ModelAttribute MyUser myUser) {
        return "register";
    }
    @RequestMapping("/checkUname")
```

```
@ResponseBody
public String checkUname(@RequestBody MyUser myUser) {
    return userService.checkUname(myUser);
}
@RequestMapping("/register")
public String register(@ModelAttribute MyUser myUser, Model model) {
    return userService.register(myUser);
}
@RequestMapping("/login")
 public String login (@ ModelAttribute MyUser myUser, Model model,
                    HttpSession session) {
    return userService.login(myUser, model, session);
}
}
```

14.6.3　Service 实现

与用户相关的 Service 接口和实现类分别为 UserService 和 UserServiceImpl。控制器获取一个请求后，需要调用 Service 层中业务处理方法，在 Service 层中需要调用 Dao 层。所以，Service 层是控制器层和 Dao 层的桥梁。

UserService 接口的代码如下：

```
package service;
import javax.servlet.http.HttpSession;
import org.springframework.ui.Model;
import model.MyUser;
public interface UserService {
    public String checkUname(MyUser myUser);
    public String register(MyUser myUser);
    public String login(MyUser myUser, Model model, HttpSession session);
}
```

UserServiceImpl 实现类的代码如下：

```
package service;
import java.util.List;
import javax.servlet.http.HttpSession;
import org.springframework.beans.factory.annotation.Autowired;
import org.springframework.stereotype.Service;
import org.springframework.ui.Model;
import dao.UserDao;
import model.MyUser;
import po.MyUserTable;
import util.MD5Util;
@Service
```

```java
public class UserServiceImpl implements UserService{
    @Autowired
    private UserDao userDao;
    /***
     * 检查用户名是否可用
     */
    @Override
    public String checkUname(MyUser myUser) {
        List<MyUserTable>userList =userDao.selectByUname(myUser);
        if(userList.size() >0)
            return "no";
        return "ok";
    }
    /**
     * 实现注册功能
     */
    @Override
    public String register(MyUser myUser) {
        //将明文变成密文
        myUser.setUpwd(MD5Util.MD5(myUser.getUpwd()));
        if(userDao.register(myUser) >0)
            return "login";
        return "register";
    }
    /**
     * 实现登录功能
     */
    @Override
    public String login(MyUser myUser, Model model, HttpSession session) {
        //ValidateCodeController中的rand
        String code =(String)session.getAttribute("rand");
        if(!code.equalsIgnoreCase(myUser.getCode())) {
            model.addAttribute("errorMessage", "验证码错误!");
            return "login";
        }else {
            //将明文变成密文
            myUser.setUpwd(MD5Util.MD5(myUser.getUpwd()));
            List<MyUserTable>list =userDao.login(myUser);
            if(list.size() >0){
                session.setAttribute("userLogin", list.get(0));
                return " redirect:/card/selectAllCardsByPage? currentPage =
            1&act=select";
            }else {
                model.addAttribute("errorMessage", "用户名或密码错误!");
```

```
            return "login";
        }
    }
}
```

14.6.4　Dao 实现

Dao 层是数据访问层，即 @Repository 注解的数据操作接口（接口中的方法与 SQL 映射文件中元素的 id 对应），与用户相关的数据访问层为 UserDao，具体代码如下：

```
package dao;
import java.util.List;
import org.springframework.stereotype.Repository;
import model.MyUser;
import po.MyUserTable;
@Repository
public interface UserDao {
    public List<MyUserTable> selectByUname(MyUser myUser);
    public int register(MyUser myUser);
    public List<MyUserTable> login(MyUser myUser);
}
```

14.6.5　SQL 映射文件

SQL 映射文件的 namespace 属性与数据操作接口对应。与用户相关的 SQL 映射文件是 UserTableMapper.xml（位于 mybatis 包中），具体代码如下：

```
<?xml version="1.0" encoding="UTF-8"?>
<!DOCTYPE mapper PUBLIC "-//mybatis.org//DTD Mapper 3.0//EN"
"http://mybatis.org/dtd/mybatis-3-mapper.dtd">
<mapper namespace="dao.UserDao">
    <select id="selectByUname" resultType="MyUserTable" parameterType="MyUser">
        select * from usertable where uname=#{uname}
    </select>
    <insert id="register" parameterType="MyUser">
        insert into usertable (id,uname,upwd) values(null,#{uname},#{upwd})
    </insert>
    <select id="login" parameterType="MyUser" resultType="MyUserTable">
        select * from usertable where uname=#{uname} and upwd=#{upwd}
    </select>
</mapper>
```

14.6.6 注册

在登录页面 login.jsp 中单击"注册"链接,打开注册页面 register.jsp,效果如图 14.10 所示。

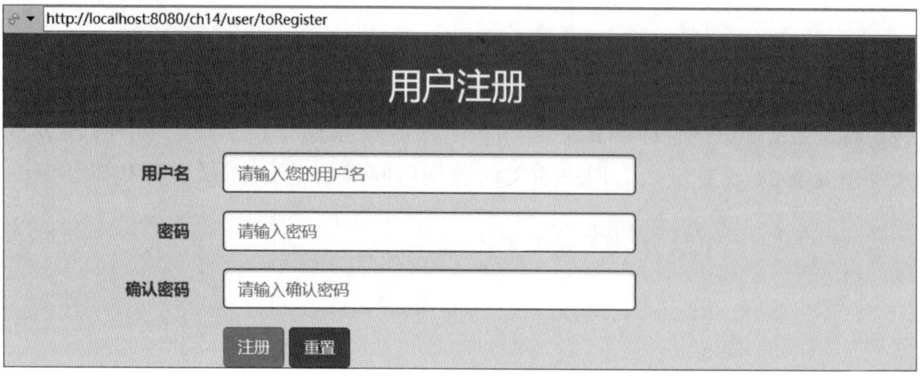

图 14.10 注册页面

在图 14.10 所示的注册页面中输入"姓名"后,系统将通过 Ajax 提交 user/checkUname,请求检测"姓名"是否可用。输入合法的用户信息后,单击"注册"按钮,实现注册功能。

register.jsp 的代码如下:

```
<%@page language="java" contentType="text/html; charset=UTF-8" pageEncoding="UTF-8"%>
<%@taglib prefix="form" uri="http://www.springframework.org/tags/form" %>
<%
    String path = request.getContextPath();
    String basePath = request.getScheme() +"://" + request.getServerName() + ":" + request.getServerPort() +path +"/";
%>
<!DOCTYPE html>
<html>
<head>
<base href="<%=basePath%>">
<meta charset="UTF-8">
<title>注册页面</title>
<link href="static/css/bootstrap.min.css" rel="stylesheet">
<script type="text/javascript" src="static/js/jquery.min.js"></script>
<script type="text/javascript">
function checkUname() {
    //获取输入的值 pname 为 id
    var uname =$("#uname").val();
    $.ajax({
```

```
            //发送请求的 URL 字符串
            url : "user/checkUname",
            //请求类型
            type : "post",
            //定义发送请求的数据格式为 JSON 字符串
            contentType : "application/json",
            //data 表示发送的数据
            data : JSON.stringify({uname:uname}),
            //成功响应的结果
            success : function(obj){//obj 响应数据
                if(obj =="no"){
                    $("#isExit").html("<font color=red size=5>×</font>");
                    alert("用户已存在,请修改!");
                }else{
                    $("#isExit").html("<font color=green size=5>√</font>");
                    alert("用户可用");
                }
            },
            //请求出错
            error:function(){
                alert("数据发送失败");
            }
        });
    }
</script>
</head>
<body class="bg-info">
    <div class="container">
        <div class="bg-primary" style="width:70%; height: 80px;padding-top:
    5px;">
            <h2 align="center">用户注册</h2>
        </div>
        <br>
<form:form action="user/register" method="post" cssClass="form-horizontal"
modelAttribute="myUser">
            <div class="form-group has-success">
                <label class="col-sm-2 col-md-2 control-label">用户名</label>
                <div class="col-sm-4 col-md-4">
                <table style="width:100%">
                    <tr>
                        <td>
    <form:input cssClass="form-control" id="uname" onblur="checkUname()"
placeholder="请输入您的用户名" path="uname"/>
                        </td>
```

```
                    <td>
                        <span id="isExit"></span>
                    </td>
                </tr>
            </table>
        </div>
    </div>
    <div class="form-group has-success">
        <label class="col-sm-2 col-md-2 control-label">密码</label>
        <div class="col-sm-4 col-md-4">
            <form:password path="upwd" cssClass="form-control" placeholder="请输入密码"/>
        </div>
    </div>
    <div class="form-group has-success">
        <label class="col-sm-2 col-md-2 control-label">确认密码
        </label>
        <div class="col-sm-4 col-md-4">
            <form:password path="reupwd" cssClass="form-control" placeholder="请输入确认密码"/>
        </div>
    </div>
    <div class="form-group">
        <div class="col-sm-offset-2 col-sm-10">
            <button type="submit" class="btn btn-success">注册
            </button>
            <button type="reset" class="btn btn-primary">重置</button>
        </div>
    </div>
</form:form>
</div>
</body>
</html>
```

14.6.7 登录

打开浏览器,通过地址 http://localhost:8080/ch14 打开登录页面 login.jsp,效果如图 14.11 所示。

用户输入姓名、密码和验证码后,系统将对姓名、密码和验证码进行验证。如果姓名、密码和验证码同时正确,则登录成功,将用户信息保存到 session 对象,并进入系统管理主页面(selectCards.jsp);如果输入有误,则提示错误。

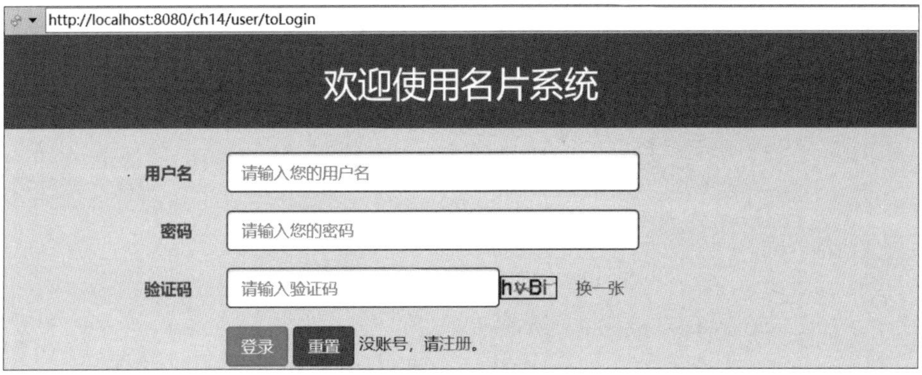

图 14.11 登录界面

login.jsp 的代码如下：

```jsp
<%@page language="java" contentType="text/html; charset=UTF-8" pageEncoding="UTF-8"%>
<%@taglib prefix="form" uri="http://www.springframework.org/tags/form" %>
<%
    String path = request.getContextPath();
    String basePath = request.getScheme() +"://" +request.getServerName() +":" +request.getServerPort() +path +"/";
%>
<!DOCTYPE html>
<html>
<head>
<base href="<%=basePath%>">
<meta charset="UTF-8">
<title>登录页面</title>
<link href="static/css/bootstrap.min.css" rel="stylesheet">
<script type="text/javascript">
    function refreshCode(){
        document.getElementById("mycode").src = " validateCode? t =" + Math.random();
    }
</script>
</head>
<body class="bg-info">
    <div class="container">
        <div class="bg-primary" style="width:70%; height: 80px;padding-top: 5px;">
            <h2 align="center">欢迎使用名片系统</h2>
        </div>
        <br>
```

```html
<form:form action="user/login" modelAttribute="myUser" method="post" cssClass="form-horizontal">
    <div class="form-group has-success">
        <label class="col-sm-2 col-md-2 control-label">用户名</label>
        <div class="col-sm-4 col-md-4">
            <form:input cssClass="form-control" placeholder="请输入您的用户名" path="uname"/>
        </div>
    </div>
    <div class="form-group has-success">
        <label class="col-sm-2 col-md-2 control-label">密码</label>
        <div class="col-sm-4 col-md-4">
            <form:password cssClass="form-control" placeholder="请输入您的密码" path="upwd"/>
        </div>
    </div>
    <div class="form-group has-success">
        <label class="col-sm-2 col-md-2 control-label">验证码
        </label>
        <div class="col-sm-4 col-md-4">
            <table style="width: 100%">
                <tr>
                    <td><form:input cssClass="form-control" placeholder="请输入验证码" path="code"/></td>
                    <td>
                        <img src="validateCode" id="mycode">
                    </td>
                    <td>
                        <a href="javascript:refreshCode()">换一张
                        </a>
                    </td>
                </tr>
            </table>
        </div>
    </div>
    <div class="form-group">
        <div class="col-sm-offset-2 col-sm-10">
            <button type="submit" class="btn btn-success">登录</button>
            <button type="reset" class="btn btn-primary">重置</button>
            没账号,请<a href="user/toRegister">注册</a>。
        </div>
    </div>
    ${errorMessage}
</form:form>
```

```
        </div>
    </body>
</html>
```

14.6.8 修改密码

单击主页面中的"修改密码"菜单,打开密码修改页面 updatePwd.jsp。密码修好页面效果如图 14.12 所示。

图 14.12 密码修改页面

updatePwd.jsp 的代码如下：

```
<%@ page language="java" contentType="text/html; charset=UTF-8"
    pageEncoding="UTF-8"%>
<%@taglib prefix="form" uri="http://www.springframework.org/tags/form" %>
<%
    String path = request.getContextPath();
    String basePath = request.getScheme() + "://" + request.getServerName() +
    ":" + request.getServerPort() + path + "/";
%>
<!DOCTYPE html>
<html>
<head>
<base href="<%=basePath%>">
<meta charset="UTF-8">
<title>修改密码页面</title>
<link href="static/css/bootstrap.min.css" rel="stylesheet">
</head>
<body class="bg-info">
    <div class="container">
        <div class="bg-primary" style="width:70%; height: 80px;padding-
    top: 5px;">
            <h2 align="center">密码修改</h2>
        </div>
```

```html
        <br>
        <form:form action="card/updatePwd" method="post" cssClass="form-
    horizontal" modelAttribute="myuser">
            <div class="form-group has-success">
                <label class="col-sm-2 col-md-2 control-label">用户名
                </label>
                <div class="col-sm-4 col-md-4">
                    <form:input  cssClass =" form - control "  path =" uname "
    disabled="true"/>
                    <form:hidden path="id"/>
                </div>
            </div>
            <div class="form-group has-success">
                <label class="col-sm-2 col-md-2 control-label">新密码
                 </label>
                <div class="col-sm-4 col-md-4">
                 <form:password path="upwd" cssClass="form-control" placeholder=
    "请输入新密码"/>
                </div>
             </div>
            <div class="form-group">
                <div class="col-sm-offset-2 col-sm-10">
                    <button type="submit" class="btn btn-success">修改
                    </button>
                    <button type="reset" class="btn btn-primary">重置
                    </button>
                </div>
            </div>
         </form:form>
      </div>
   </body>
   </html>
```

在图 14.12 中输入"新密码"后单击"修改"按钮,将请求通过 card/updatePwd 提交给控制器类。根据请求路径找到对应控制器类 CardController(14.5.2 节)的 updatePwd 方法,处理密码修改请求。这里找控制器类 CardController 处理密码修改,是因为用户必须登录成功后才能修改密码。

14.6.9 安全退出

在管理主页面中单击"安全退出"菜单,将返回登录页面。"安全退出"超链接的目标地址是一个请求 card/loginOut,找到控制器类 CardController(14.5.2 节)的对应处理方法 loginOut。这里找控制器类 CardController 处理安全退出,是因为用户必须登录成功后才能安全退出。

14.7 小　结

本章讲述了名片管理系统的设计与实现。通过本章的学习,读者不仅掌握 Spring MVC＋MyBatis 应用开发的流程、方法和技术,还应该熟悉名片管理的业务需求、设计以及实现。

习　题　14

1. 在名片管理系统中,用户是如何控制登录权限的?
2. 在名片管理系统中,安全退出功能的程序做了什么工作?

参 考 文 献

[1] 杨开振.Java EE 互联网轻量级框架整合开发：SSM 框架（Spring MVC＋Spring＋MyBatis）和 Redis 实现[M].北京：电子工业出版社,2017.

[2] 黑马程序员.Java EE 企业级应用开发程序（Spring＋Spring MVC＋MyBatis）[M].北京：人民邮电出版社,2017.

[3] 陈恒.Java EE 框架整合开发入门到实践：Spring＋Spring MVC＋MyBatis（微课版）[M].北京：清华大学出版社,2018.

[4] 王耀.深入理解 Spring MVC 源代码：从原理分析到实战应用[M].北京：中国水利水电出版社,2019.

图书资源支持

感谢您一直以来对清华版图书的支持和爱护。为了配合本书的使用,本书提供配套的资源,有需求的读者请扫描下方的"书圈"微信公众号二维码,在图书专区下载,也可以拨打电话或发送电子邮件咨询。

如果您在使用本书的过程中遇到了什么问题,或者有相关图书出版计划,也请您发邮件告诉我们,以便我们更好地为您服务。

我们的联系方式:

地　　址: 北京市海淀区双清路学研大厦 A 座 701

邮　　编: 100084

电　　话: 010-83470236　　010-83470237

资源下载: http://www.tup.com.cn

客服邮箱: 2301891038@qq.com

QQ: 2301891038(请写明您的单位和姓名)

用微信扫一扫右边的二维码,即可关注清华大学出版社公众号"书圈"。

资源下载、样书申请

书圈

扫一扫,获取最新目录

课程直播